HARTFORD PUBLIC LIBRARY
Alexandria, Ohio

BUILDING YOUR OWN HOME

BUILDING YOUR OWN HOME

Fred Yeck

ARCO PUBLISHING COMPANY, INC.
219 Park Avenue South, New York, N.Y. 10003

Published 1975 by Arco Publishing Company, Inc.
219 Park Avenue South, New York, N.Y. 10003
by arrangement with Home Builders Plans

Copyright © 1975 by Fred Yeck

All rights reserved. No part of this book may be reproduced, by any means, without permission in writing from the publisher, except by a reviewer who wishes to quote brief excerpts in connection with a review in a magazine or newspaper.

Library of Congress Catalog Card Number 74-16918
ISBN 0-668-03656-7
Printed in the United States of America

Contents

1. Should You Build Your Own House? ... 1
2. Buying Land ... 3
3. Planning Your New Home ... 5
4. House Plans ... 8
5. Financing ... 12
6. Permits, Regulations, and Insurance ... 14
7. Buying Materials ... 17
8. Preparation of Site ... 19
9. Excavation ... 23
10. Footings ... 26
11. Foundation and Basement ... 29
12. Waterproofing the Foundation ... 38
13. Slab Floors ... 40
14. Pier Foundation ... 44
15. Floor Framing and Subflooring ... 48
16. Wall Framing and Sheathing ... 54
17. Roof Framing and Sheathing ... 66
18. Roofing ... 76
19. Windows ... 79
20. Doors ... 83
21. Flashings, Vents, and Gutters ... 86
22. Exterior Siding and Trim ... 89
23. Plumbing and Drainage ... 92
24. Septic System ... 103
25. Plumbing and Water Supply ... 107
26. Well ... 111
27. Plumbing Fixtures ... 113
28. Plumbing for Gas ... 115
29. General Heating ... 117
30. Hot Water Heating ... 119

31.	Forced Warm Air Heating	121
32.	Heating: Perimeter for Slabs	125
33.	Air Conditioning	126
34.	General Electrical Service	128
35.	Electrical Installation	134
36.	Insulation	148
37.	Wall and Ceiling Finish	150
38.	Cabinets	153
39.	Interior Trim	159
40.	Finish Floors	161
41.	Painting and Wallpapering	164
42.	Appliances	166
43.	Concrete Flatwork	168
44.	Brick	171
45.	Fireplace	174
46.	Grading, Seeding, and Sodding	176
47.	Lumber Sizes	178
48.	Nails and Screws	180
49.	Tools	181
50.	Glossary of Terms	183
51.	Specifications	193
52.	Materials List	199
Index		213

BUILDING YOUR OWN HOME

1. Should You Build Your Own House?

A new house is the largest expenditure that most people will ever make. Therefore, it should be a satisfying and enjoyable experience. The best way to determine if you should build your own house is to examine what kind of a person you are. You can do this by answering the following questions:

Do you have any building or home repair experience?
Have you ever had any hobbies which required any type of building or planning?
Do you have any special technical skills?
Have you had experience building anything in connection with your occupation?
Have you ever sketched house plans or thought about what kind of house you would like to build?
Do you enjoy shopping in lumber yards and hardware stores?
Do you own tools and enjoy using them?
Have you had experience taking on large jobs?
Do you usually complete any job you start?
Do you know anyone who has built a house and can explain the difficulties to you?
Do you really want to build a house?

Some other considerations are:
Can you afford to build a house? Any loan company can help you decide this. Never build a house yourself if your only reason is to save money.
Do you have the time required? Building a home will require weekends and evenings for about a year, plus a couple of weeks' vacation.
Do local laws allow you to build your own home? Almost all

building codes allow an owner to build his own home as long as the code is followed.

Are there good local sources of all building materials you will need?

Do you have someone to help do the heavy work such as pouring cement or assembling roof rafters? This work actually requires only about three or four days.

Do you have the necessary transportation to get materials? Most dealers will deliver but cash-and-carry dealers offer lower prices.

The best rule to remember is that if you think you can build a house, chances are you can. It's a big job, but not nearly as big as it might seem in the beginning. Consider each small job separately because that is the way a house is built, one job at a time.

Try to find someone who has built his own home—he will be more than eager to tell you all about it. You will get a lot of encouragement and learn a lot, too. Be careful when talking to people who have had a home built by a contractor and did some of the work themselves—in some cases they were interested only in saving money and did not really enjoy doing the work. Above all, don't be discouraged by people who have never tried it.

2. Buying Land

Even if you already own land, you should go through the following steps before building:

Inspect the lot.

   Is the lot close enough to schools, churches, shopping areas, and transportation?
   Is there police and fire protection?
   Is the lot large enough? Does it have adequate road frontage?
   Is there proper drainage?
   Are the roads adequate?
   Does the site fit the kind of house you want to build?
   Are homes in the area comparable to the one you plan to build?

Ask the neighbors some questions.

   Do basements in the area ever fill with water?
   If land is near rivers or lakes, does it ever flood?
   Are there any offensive businesses or conditions in the area?
   Are the schools good?
   Is the subsoil rocky? Ask an excavator who knows the area or get the permission of the owner to drill test holes with a post hole digger.

Purchase a copy of the local building code from the city clerk or county building and zoning department, whichever applies.

   Does the code allow you to build your own house? Almost all codes do, although you will have to abide by the code and be prepared to pass inspection of the work.

Some areas have very backward building codes. Read the code and note any variations in the plans. Will any variations make it difficult to build the house?

Ask the utility companies and the subdividers of the land the following questions. Don't take the owner's word for it.

Is water available at the lot line? If not, what will it cost to get water?

Are there city sewers at the lot line? What is the fee for connecting to them? What is the annual rental fee?

Are there electric lines at the lot line? If not, what will it cost to have lines brought to the lot line?

Similarly, is gas available at the lot line? If not, what will it cost to have it brought to the lot line?

Is telephone service available?

If all the utilities are not available, substitutes will have to be found. Ask local contractors and then double-check by asking neighbors.

How much will a well cost?

Is the ground suitable for a septic tank? What will it cost?

Is the cost of electric heat, in place of gas, practical? Is the cost of oil or bottled gas practical? For an impartial opinion, ask a contractor who sells both.

Before buying a lot, get an attorney to handle the purchase.

Ask the attorney what his fee will be.
Is a clear title available?
Are there any liens or back taxes on the property?
Get the attorney's opinion of the property.
Is a survey or any soil test recommended?

When you are ready to buy, ask local building and loan associations for their opinions of the property.

3. Planning Your New Home

The following steps will help you plan the kind of home you should build. Check your local building code for other considerations.

Plot planning

The house should not cover more than 30 per cent of the lot.

The house should be at least 15 feet from the front or rear lot line and 5 feet from the side lot lines.

Locate the house for the best view and protection from cold north or west winds and hot south and west sun.

Position the house to avoid any trees that you want to preserve.

Large glass areas should be protected from the wind and sun. The west side is usually the hottest and also the windiest.

A terrace or patio on the east or north side provides shade for evening dining outside.

Bedrooms on the north will be cooler in the summer. Also, you will be able to sleep late without the sun shining in.

An east kitchen is cheerful in the morning and cool the rest of the day. A garage on the south end can have a greenhouse attached.

The garage should be located to provide easy access to the street and, in cold climates, to avoid the possibility of snow drifts blocking it.

Bedroom windows should be situated so that they do not open directly toward neighbors' windows.

Number and use of rooms

Bedrooms: Unless you are building in an area of retirement or vacation homes, you should build at least a three-bedroom home.

The resale value and ease of selling are much better with a three-bedroom house. Retirement homes should have at least two bedrooms and vacation homes can have any number of bedrooms.

Bathrooms: A three- or four-bedroom house should have two baths or one bath and a half-bath. If the house has a basement and cost is to be low, the half-bath may be in the basement.

Kitchen: The kitchen must have an eating area if the house has no dining room. If you have a large family, a family room–kitchen combination will give you more room for dining and can also be used for other purposes.

Dining room: A dining room is preferred if you enjoy formal dining or entertaining. A true dining room is always separated from the kitchen by a wall.

Family room: If you have children, a family room provides a separate and informal area for the family to use, and you still have the living room for adults or guests. Family rooms are often connected to the kitchen and have direct access to the backyard or patio.

Closets: All bedrooms should have at least 3 feet of closet space; master bedrooms should have 6 feet. There should also be a linen closet and a coat closet. Houses without a basement should have a large storage closet, too.

Basement: A basement provides much extra space at low cost. If the house is built on a hill, a basement with a walk-out door and large windows is especially nice.

Utility room: If a house has no basement or garage, it should have a utility room, which can be used as a laundry room. A utility room is not needed for the furnace unless the house has no attic or crawl space.

Entry hall: An entry hall is very convenient unless the house is small. The entry gives the living room much more privacy and cuts down on drafts caused by people going in and out in cold weather.

Other rooms: A laundry area can be used in houses with a basement if you prefer having the laundry upstairs. A kitchen with a dinette or breakfast room can be used for family meals in a house which has a formal dining room. Small berth-type bedrooms or large dormatory-type bedrooms can be used if you have a large family. In a large house, you may want a study or den. An extra half-bath, located next to the garage and sometimes called a "Mud room," is handy in larger homes. A half-bath located near the front entry is sometimes called a "Powder room." Large bedrooms can have a dressing room between the bedroom and attached bath.

Walk-in closets are also used where space permits, and since only one door is used, this leaves more wall space in the bedroom.

Two-stories and split-levels

Two-story houses provide the lowest-cost construction for larger homes.

Split-levels offer more usable space by providing an extra area which is partially below grade. However, constructing one is more difficult for an amateur builder.

A raised-ranch is a single-story house with a basement, but the basement floor is only 3 or 4 feet below grade, and a ground-level entry is used. If the lower floor is counted as living space, this offers the lowest-cost construction for its space.

Garage, carport and outdoor storage

If a house has no basement, an attached garage should be considered. An attached garage costs less than a separate garage and is kept warmer in cold weather by heat from the house. A two-car garage is only a little more expensive than a single-car garage and is much more convenient. A two-car garage is preferred with a three-bedroom house and will greatly increase the resale value.

A carport is practical in warm climates. It will cost less than a garage since no walls or continuous footings are required.

Outdoor storage space should be provided. If a house has no basement or garage, an outdoor storage shed will definitely be needed. A garage may be used for storage, and, in this case, a dividing wall for separate storage space will eliminate a cluttered appearance. A storage space can also be built into a carport.

4. House Plans

When people refer to house plans, they are usually thinking of the small floor plan drawings which are found in newspapers and house magazines. These are very helpful in determining what type of home you want to build. There are also a number of businesses that sell booklets of house plans. These can be found wherever magazines are sold, and are also advertised in house magazines.

The actual house plans, of course, are the working drawings which contain all the dimensional information needed to build the house. These usually consist of the floor plan, a foundation plan, four elevations (or views) of the house from all four sides, a cross section showing construction details, and interior details showing stairways, bathroom cabinets or tile, kitchen cabinets, and so forth. You can order these working drawings from any of the sources mentioned above for about $25 to $50 per set. This usually includes a materials list and specifications outline. It is also possible to hire a drafting firm to draw plans to your specifications. These will cost from about $150 to $300.

An excellent source of house plans is Home Planners, Inc., 16310 Grand River Avenue, Detroit, Michigan 48227. They have over 600 plans, and they are designed and dimensioned in a way which makes them especially easy for an amateur builder. If you would rather design your own home, Home Builders Plans, 136 Scott Drive, Dundee, Illinois 60118, will draw the plans for you from your sketches or pictures, and will help you with any problems you may have with the design.

Before starting any work, study your plans, and read the directions completely through. Be sure you understand every step. Most of the work can be figured out as you progress, but be especially careful with the concrete work. It's hard to correct if you make a mistake!

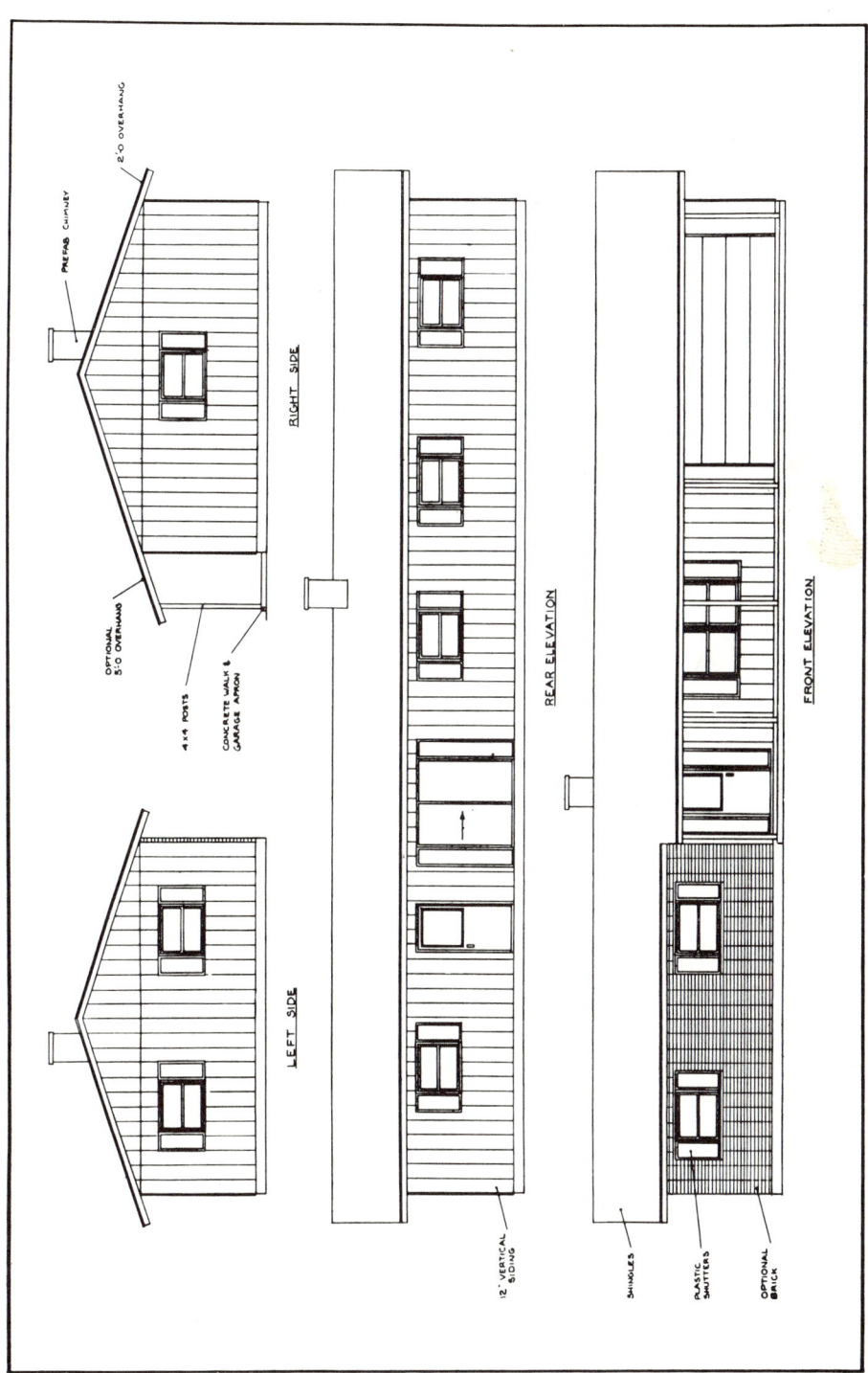

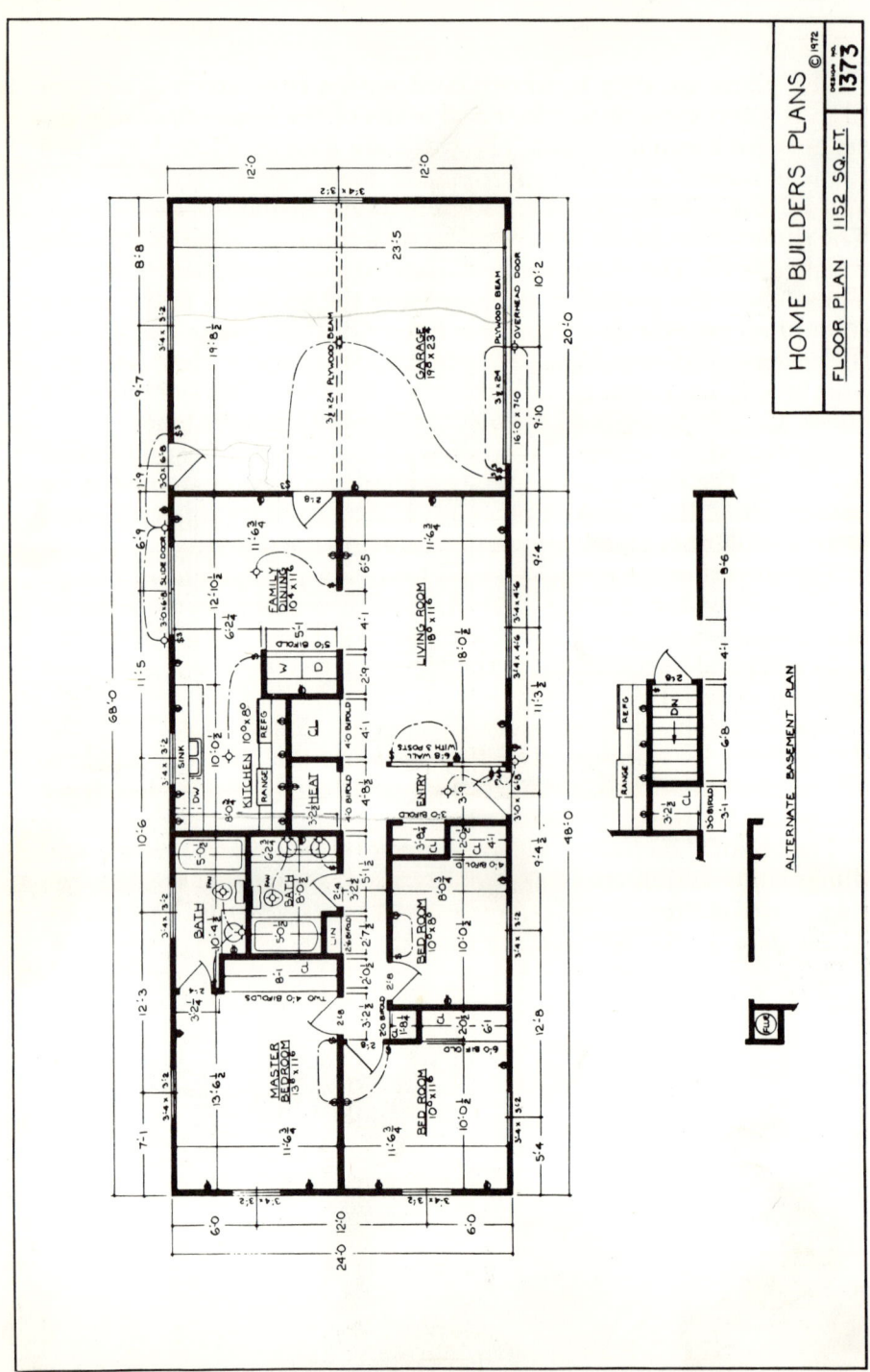

The plans are easy to understand with a little study. The floor plan is a top view of the floor and walls of the house drawn to the scale ¼ inch equals 1 foot. The walls are usually shaded-in. Each room is named with the approximate room size under the name, and all the interior walls are dimensioned inside the house plan. The dimensions given are from the wall framing, and not from the finished wall. The exterior dimensions are shown outside the plan. These show the overall dimensions and the locations to the center lines of all exterior doors and windows. On some plans, the sizes of the doors and windows are shown; on others they are just numbered, and the materials list gives the sizes. The floor plan also shows the locations of all the electrical outlets and switches.

The foundation plan is a top view of the foundation or basement walls. All the dimensions are the actual dimensions required for constructing the foundation. Gravel fill, reinforcing wire, steel beams, and other materials are indicated on the drawing.

The elevation drawings show all four views of the house, and are used mainly for getting permits and financing. Some exterior materials such as siding and roofing are indicated on this drawing.

The typical sections show the sizes for rafters, joists, and most framing materials.

The interior details show stairway dimensions, and the dimensions for cabinets and tile work for the bath and kitchen. These drawings usually just have overall dimensions, and leave the details for the carpenter.

The plans included in this book show many views and dimensions which are not generally used on house plans because they are common to most houses and are known by carpenters. These details will help you interpret the plans.

5. Financing

Obtaining financing may well be the hardest part of building a home. Some lenders may discourage you from building if they don't have money to lend. Others may be more willing to do business, so don't give up easily if money seems hard to come by. The following steps will enable you to get a loan if you qualify.

Estimate the cost of the house you plan to build and take the plan and specifications to every bank and building and loan association in your area. Explain what you want to do. They will ask about your savings, income and debts, and will advise you on the chances of obtaining financing. You will probably be required to have enough money to cover the cost of having the house completed in case you become ill or injured and cannot finish it yourself. For this reason the down payment will be the same as if you were buying the house and not doing any work yourself. However, when you complete the house, the amount borrowed may cover most of the actual costs, so your down payment money can be used to reduce the loan, or can be used to buy furniture or other necessities.

If financing is readily available, you can shop for a low interest rate or low down payment. Some banks have an initial charge of 1 per cent or 2 per cent of the loan to get a lower interest rate. This can be to your advantage if you plan to keep the loan for a long time. When you have found financing, you will have to get the loan approved as follows:

Take your specifications and plans to at least two building contractors and get rough estimates on the entire finished house. Be sure that the contractor tells you if he charges for estimates and how much. Let him know that you are getting a rough estimate to determine loan value only. Sometimes a lending institution will make its own estimate of how much the house will cost.

Fill out the materials list and take it to a building supply dealer and get an estimate on all the materials. Many building supply dealers now carry all materials required, including plumbing, heating, and electrical supplies. Try to pick the dealer who has the best prices. (See *Buying Materials.*)

Have your attorney ready to read the mortgage agreement. Find out how much he will charge.

Take your specifications, estimate, and plans to the lending institution. Ordinarily only the floor plan and the outside elevations are required. It will usually take about a week to get the loan approved. Sign the loan agreement only in the presence of your attorney.

The money will usually be paid out at certain intervals set by the lenders. Sometimes they will pay the bills directly. They will also inspect the progress of the house.

Buy any insurance the lenders may require.

6. Permits, Regulations, and Insurance

Most plans are drawn to meet Federal Housing Administration requirements. They should, therefore, meet most building codes. However, some areas have very strict or antiquated building codes which do not allow certain modern methods of construction, and unless special permission is given, these codes must be obeyed.

Purchase a copy of your local building ordinance or building code. If you are building in a city or village, the city clerk will have copies of the code. If you are building outside of a city, the county code will apply—it can be purchased from the county building and zoning department. This code must be followed strictly in order to pass inspections.

Become acquainted with your local building inspector and tell him what you are planning. He can be very helpful in interpreting the building code and giving you information about building regulations and methods in your area. Building inspectors may also know shortcuts used in the area, which could save you time and money. There are sometimes changes in the code which are not yet in print. Often these allow you to use more modern methods of construction.

The building inspector will inform you of any inspections, permits, and regulations you must abide by. Be sure these regulations are met.

In some areas a plot plan must be submitted to the city or county. If this is required, it can be done easily by providing the following information:

Drawing of plot to scale of 1/16 inch equals 1 foot. Indicate the scale at the bottom of the drawing.

Legal description of property from your title for the land.

Dimensions of the plot and arrow indicating north.

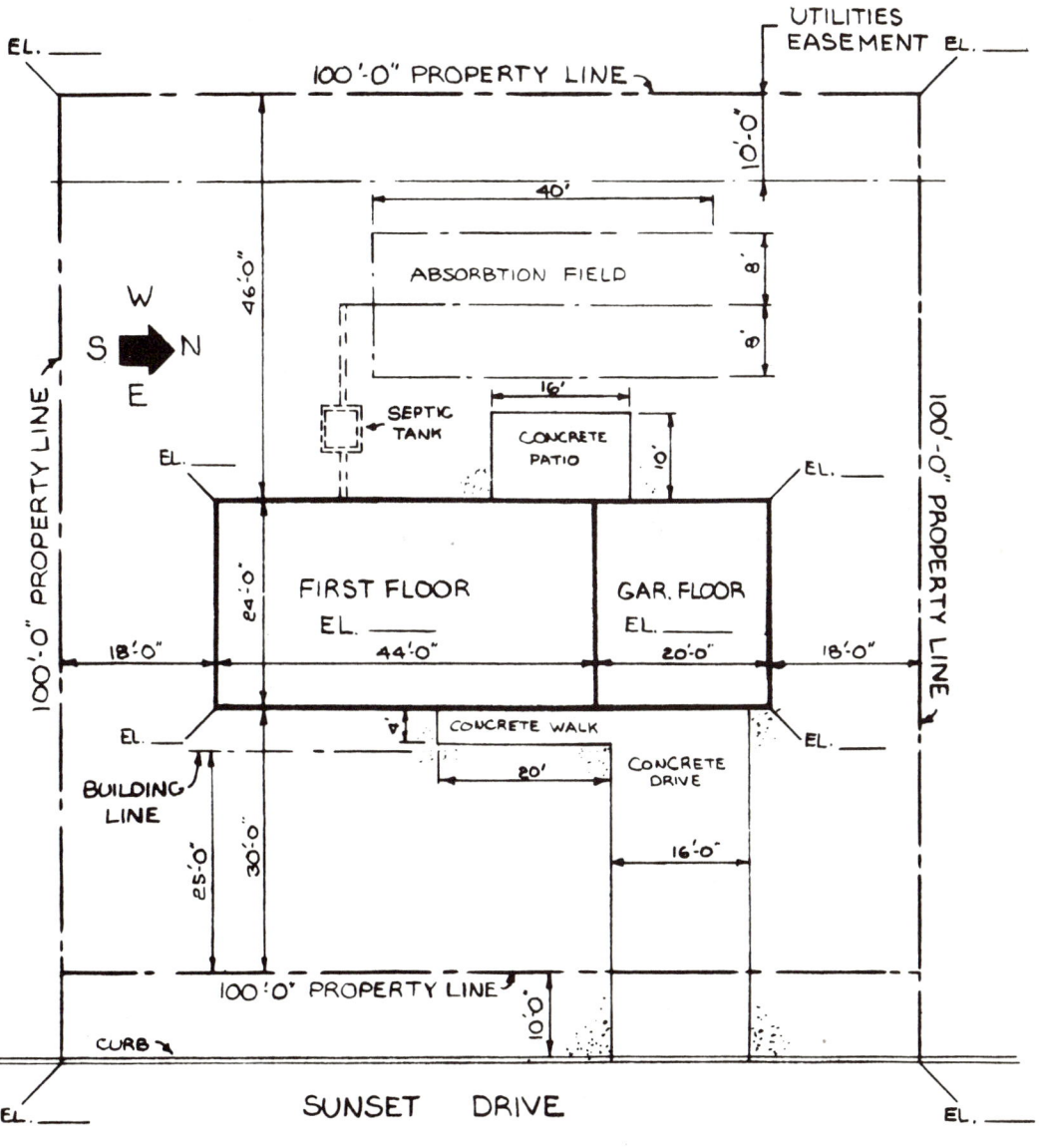

Lot 3 Block 4
OAK RIDGE - DUNDEE, ILLINOIS

SCALE 1/16"=1'-0"

PLOT PLAN

16 PERMITS, REGULATIONS, AND INSURANCE

Plan of house and garage or carport.

Dimensions of front, rear, and side yards.

Location of walks, driveways, and approaches.

Location of steps, terraces, porches, fences, and walls.

Location of easements and required setbacks. This will be indicated on your title for the land.

Location of well, septic tank, and absorption field, if used.

If a plot plan is required for a hilly or sloping plot, elevations may be required. If the city or county won't accept your plot plan without the elevations, have a surveyor add them to your finished drawing in the places marked E1.———

Before any building is started, purchase homeowner's insurance to protect yourself from being sued by anyone who is hurt on your property during the construction. This insurance should also cover the house in case of fire, storms, and so forth.

7. Buying Materials

Building materials must be purchased from a reputable dealer at the lowest cost possible. A few cents difference per article multiplied by the quantity of materials in a house can make a big difference. The best place to buy most materials is the large cash-and-carry building materials dealer. If there is one in your area and you have a truck, station wagon, or trailer, check their prices. If you don't have the transportation, you will have to buy from a dealer who will deliver the materials. Check his prices against cash-and-carry dealers, and you may find it will pay to rent a truck occasionally to get the materials from cash-and-carry dealers. This will certainly pay when buying the framing, sheathing, siding, windows, doors, and interior wall materials.

Plumbing, heating, and electrical supplies can also be purchased from cash-and-carry dealers. Most of these materials can be carried in a station wagon, but it will take quite a few trips, so look into renting a truck here also. If no cash-and-carry dealer is available, check with plumbing, heating, and electrical dealers in your area and get price quotations for the items on your list. You can also get these materials from the large mail order catalogs or stores, but the prices are usually higher, especially with shipping charges. Heating and air conditioning can be purchased as a package deal guaranteed by the dealer. This is very convenient, and has the advantage of assuring availability of repair service and parts.

Floor coverings and carpeting can be purchased from local dealers who have large selections. This is a very competitive business so shop carefully for the best price.

Discounts are available from almost all building supplies dealers except the large cash-and-carry dealers. You should be able to get a contractor's discount on everything—this amounts to 10 or

15 per cent. Explain that you are building a house and offer to buy all the materials from the one dealer if you get a good price. This may seem like a lot of effort but the savings can amount to as much as $2000 on a whole house, which is well worth it.

Start charge accounts at the places you decide to do business with. If you already have a mortgage, have them check with the lender and you will have no trouble getting credit. Try to get 30, 60, or even 90 days to pay the bills, then pay your bills promptly at the end of this period to avoid interest. Some dealers and subcontractors will let bills ride for six months or longer, which can save you a lot in interest. Never buy anything on time payments or revolving charges, as the interest rate is very high compared with a mortgage loan rate.

8. Preparation of Site

The house should be located on the site to conform to the local building codes. Also, it should be situated with some regard to the other houses in the area so it will not look out of place. Many newer areas have the houses set back at different distances from the street for variety. If the houses are all in a straight line or parallel to the street, be sure to keep your house parallel also, so it will not look crooked. Drive in temporary stakes to locate the corners of the house.

Build batter boards about 4 feet back from the stakes. Use 2 x 4's for the posts and 1 x 4's for the boards. Drive the posts firmly into the ground to eliminate any chance of their moving. Stretch strings between the boards so that they cross over the tops of the stakes. Use strong string sold especially for this purpose. Hang a line-level in the center of the string and level the boards by driving the highest posts deeper into the ground.

When all the boards are level, you are ready to lay out the exact size and shape of the house. Measure the distances between the strings with a 100-foot steel tape. Use the overall dimensions from the floor plan.

Position the front string exactly where you want the front of the house. Next, mark the location of the two front corners on this string. Measure from the lot line to the position where you want a corner. Knot a short piece of string at this point. Measure the width of the house and garage along the front string and tie on another short string. Measure back the correct distance from the front string and locate the rear string. Tie short strings to the back string in the same method used for the front. Now measure the diagonals. Move the two knots on the string equal amounts until the diagonals are exactly equal.

An easy way to make these measurements is to let the short

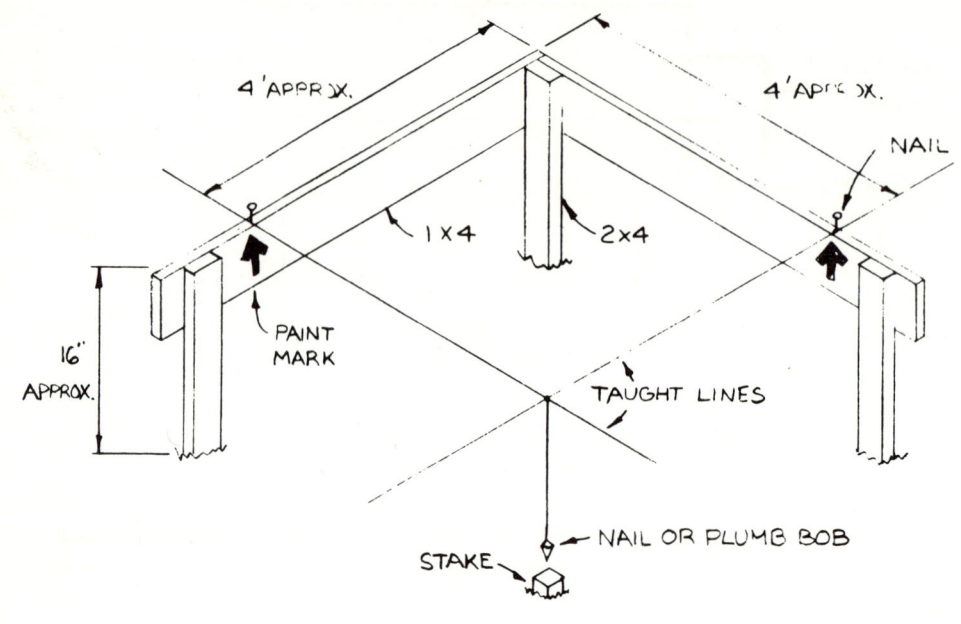

BATTER BOARDS

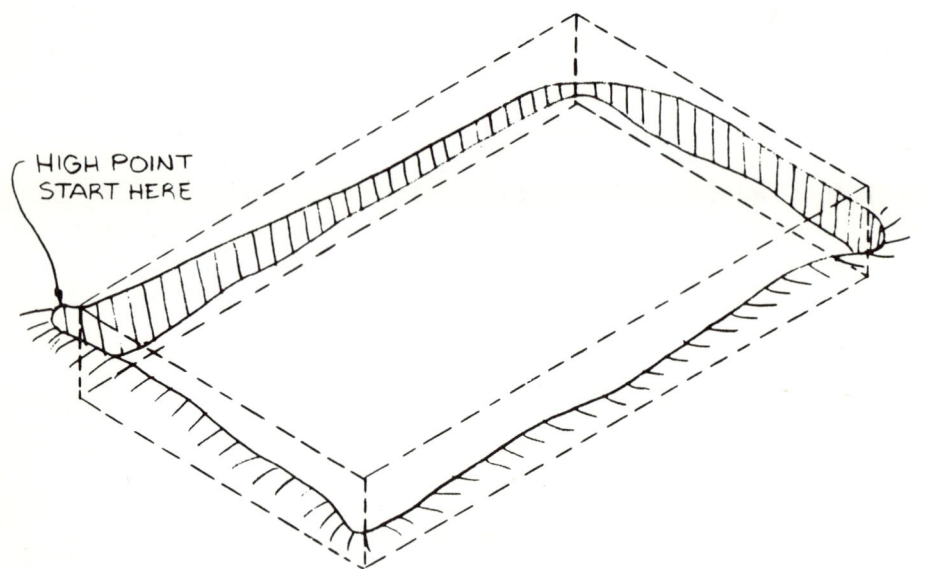

BUILDING LAYOUT - SLOPING SITE

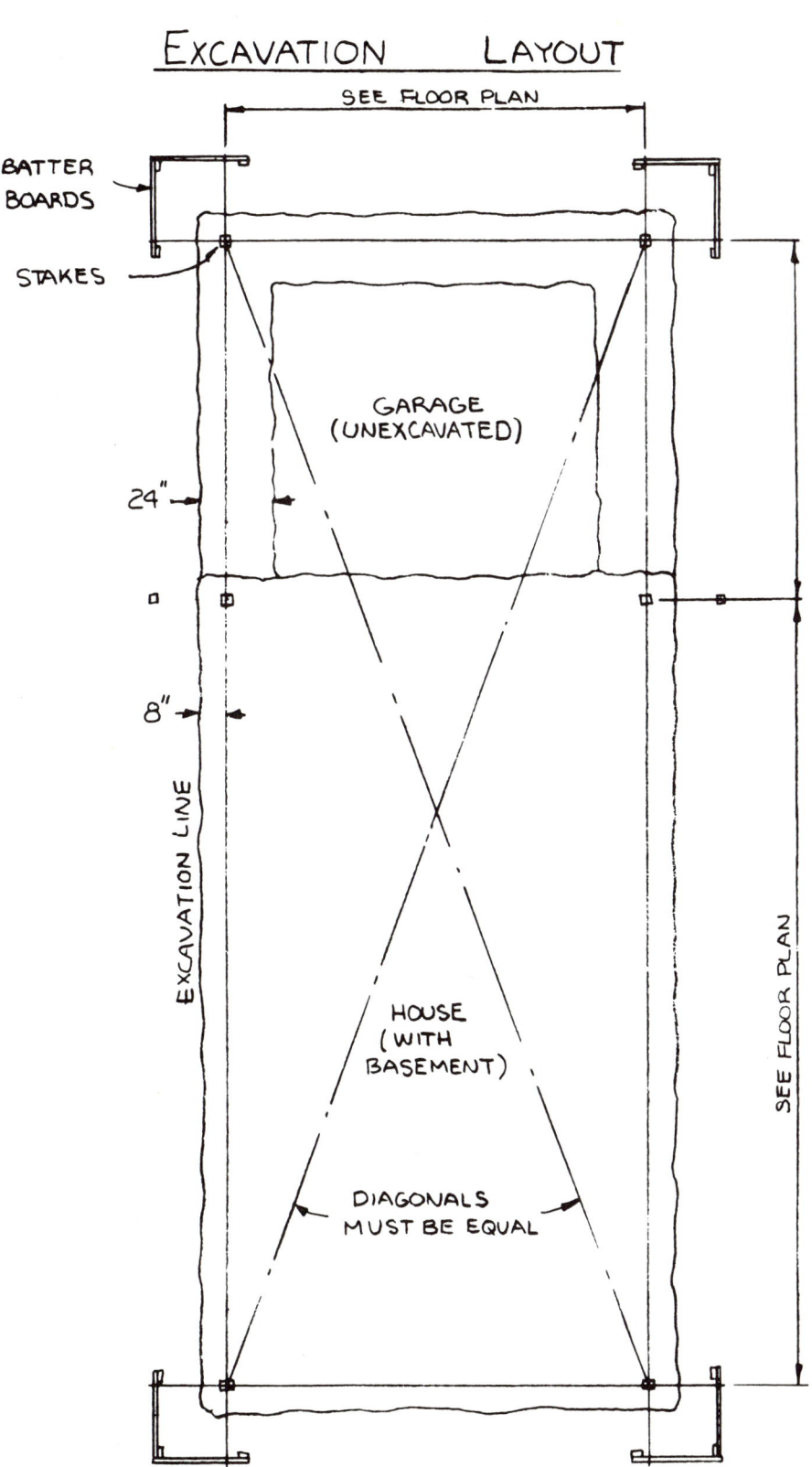

strings hang down to ground level and tie plumb bobs to the bottoms. Then have someone hold the zero point of the steel tape under one weight while you hold the other end under the next weight. Move the strings until the measurements are all correct.

Next, move the end strings until they cross exactly over the knotted strings. Carefully remeasure all the dimensions. Hammer nails into the batter boards to mark the position of each string. Paint a large mark by each nail to guide the excavator. Measure the position of the wall between the house and garage and drive in a stake about four feet outside the line. This will mark the point where the house excavation stops, so the excavator can leave the garage area unexcavated.

If the house is to be built on a post foundation, measure the location of all the posts, and mark each with a stake. Dig all the post holes by removing one stake at a time.

9. Excavation

The depth of the excavation will depend on the depth of the footings. The bottom of the footings must be below the frost line; this is regulated by the local building code. This depth must be maintained from the finished grade. If the footings must be 40 inches deep and there will be 10 inches of fill, the excavation will be 30 inches deep.

To determine the depth of fill which will be removed from the excavation, use the following method. The excavated material will usually raise the finished grade by 25 per cent of the excavation depth. The depth of excavation will then be about 80 per cent of the required footing depth. Example: If the footings must be 40 inches deep, the depth of the excavation will be 40 times 80 per cent or 32 inches. This method is for dug out crawl spaces or basements only.

If fill will be brought in from another area, subtract the depth of the fill from the depth of excavation.

If wood footing forms are used, add 6 inches to the depth of excavation. (See *Footings* to determine if wood forms will be used.)

For full basements, the depth of the footings is 84 inches, and the depth of the excavation is 67 inches.

If the house is on a slab, excavate for the footings only and figure about 2 inches of fill. If more fill is required to build the lot to the desired level, it will have to be hauled in.

For crawl spaces, 24 inches of depth below the floor joists is the minimum, but you will probably want more room to work. Footing depth must be at least as deep as the crawl space. In cold climates where footing depth is over 40 inches, you may want to excavate the crawl space to only 40 inches. This will give a 48-inch depth under the floor joists, which is more than ample. Excavate deeper

EXCAVATION

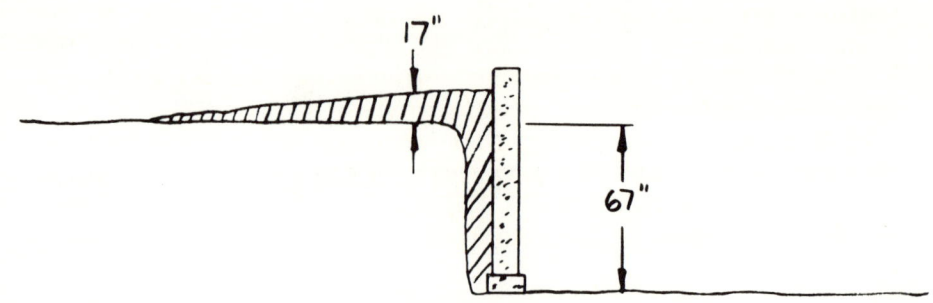

BASEMENT

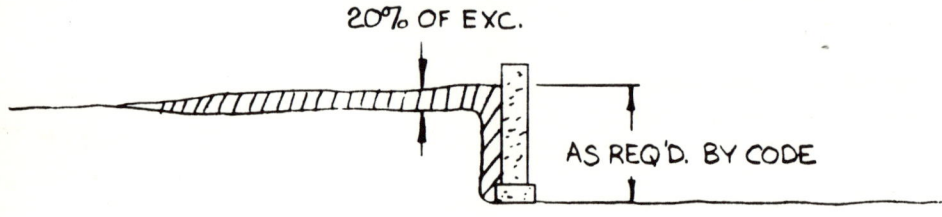

CRAWL SPACE

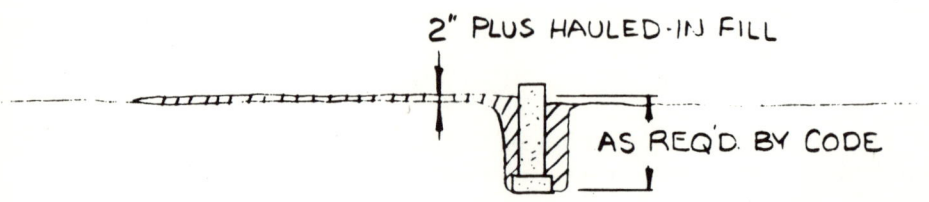

SLAB

for the footings if required by the building code.

Garage areas are excavated for the footings only. The garage footings are dug to the depth required by the code. If the foundation is to be made of concrete block, the distance from the basement footing to the garage footing must be a multiple of 8 inches, so that the tops of the foundations will line up.

Hire an excavator and tell him what you want. Give him the foundation plan, and tell him how deep you want the footing trenches and basement, if you will have one. Show him the marks on the batter boards. Of course, if the necessary equipment is available, you can do this job yourself. Footings for slabs or garages can sometimes be dug by hand.

As soon as the excavating is done, the sewer and water can be connected, if desired. (See *Plumbing and Drainage.*) An electric utility pole can also be installed so that you can use power tools. (See *General Electrical Service.*)

10. Footings

First determine the type of footings to be used. If the bottom of the excavation is very level and free of rocks, a trench form can be used. If the excavation is uneven or rocky, it may be easier to build wood forms for the footings. This should be determined before the excavation is complete so that the excavator can dig an extra 6 or 8 inches deep for the footings. Basements must always have wood forms so the footings will be above the ground, allowing space for the gravel fill required under the floor.

If the foundation is to be cement block, make sure that the dimension from the top of the basement or crawl space footing to the top of the garage footing is an exact multiple of 8 inches so that the blocks will come out even on top.

Layout of footings is done by first replacing the strings on the batter boards. At the point that the strings cross, tie a string with a plumb bob on it and let it hang down to the bottom of the excavation. Drive a stake directly under the plumb bob and drive a nail into it to locate the plumb line.

Recheck the dimensions against the floor plan, and recheck the squareness to be sure it is still right. (See *Preparation of Site*.) Level the stakes by tying a string to the nails and putting a line-level in the exact center of each string. Drive in the highest stakes until the lines are all level.

The strings indicate the outsides of the house walls. The footings are wider than the walls so they will extend outside the strings. See the foundation drawings for the correct width and depth of the footings and how far they project outside the house walls. Measure the position of the center post footings and mark with a stake. These footings are 2 feet square and at least 8 inches thick.

If trench forms are used, dig the trench to the required depth of 6

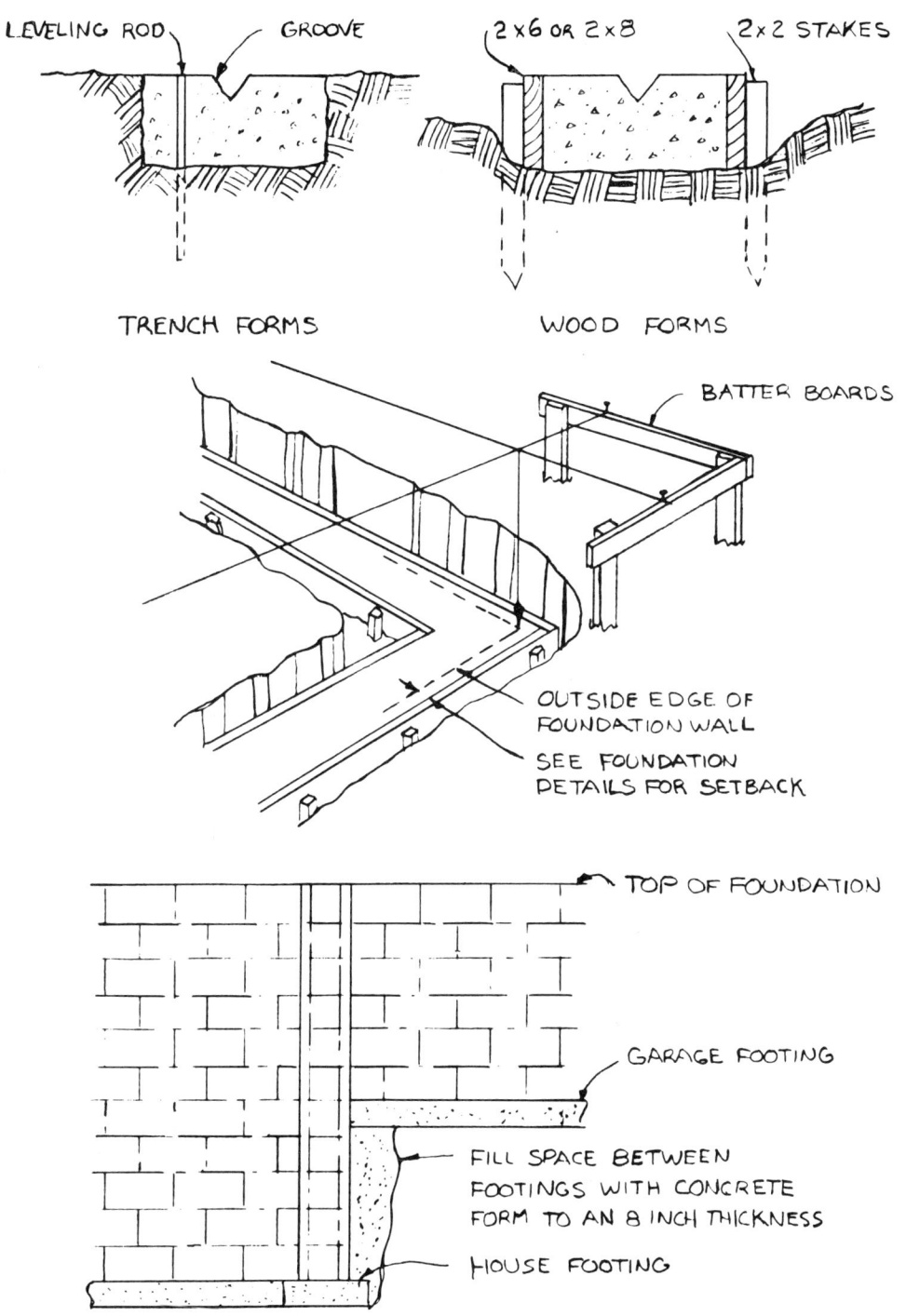

FOOTING DETAILS

or 8 inches and drive pipe or steel rods every 5 feet for leveling the trench. Wood stakes can also be used but they must be removed before the concrete is hard.

If wood forms are used, use 2 x 6 or 2 x 8 framing lumber and drive 2 x 2 stakes every 5 feet to nail the forms to.

Level the forms or trench stakes by sighting over them by eye. Be sure the corners remain level. Another way to check levelness is to select a perfectly straight 2 x 4 about 16 feet long, place it on the forms or stakes, and put a carpenter's level on it. The line-level cannot be used for this because of the sag in the line.

Caution: Do not leave any loose soil or fill under the footings.

If you plan to mix your own cement, an electric- or gasoline-engined mixer can be rented for a day. The cement is usually mixed in the proportions of 1 part cement to 2½ parts sand and 5 parts stone. Your local dealer can estimate the quantities you should buy if you give him the width and thickness of the footings and total feet of foundation. Mixed and sacked cement can also be purchased.

If you use ready mix cement, tell the supplier that you want a mix suitable for footings and tell him how many cubic yards you need. Add up the total feet of footings and multiply by the width in inches. Then muliply by the thickness in inches, and divide by 3900. This will give you cubic yards. Example: 150 feet of footings times 6 inches thick times 16 inches wide equals 14,400. Then, 14,400 divided by 3900 equals 3.7 cubic yards.

Be ready to level the footings as soon as they are poured. A 2-foot length of 2 x 4 can be used to level the concrete with the top of the forms or stakes. This is usually a one-man job. Using a trowel, cut a 1½-inch-deep groove down the center of the footing. This is to key the foundation wall into the footing.

If the foundation is to be poured concrete, the foundation contractor may want to pour the footings himself. (See *Foundation and Basement.*)

11. Foundation and Basement

Foundation walls for basements should be 88 inches high. This is 11 courses of concrete block. These walls should extend 8 inches above the finished grade. The foundation wall for crawl spaces or slabs should also extend 8 inches above the finished grade. (See *Excavation*.) If concrete block is used, the wall height must be in multiples of 8 inches.

See the foundation drawings for details of building the walls. See the foundation plan for the correct dimensions.

To estimate the number of blocks needed, multiply the total feet of foundation times the number of courses, times .75. Example: 150 feet times 6 courses times .75 equals **675** blocks. For slabs or brick veneer houses, the top course is a different size so figure this separately. Each corner will require corner blocks. Multiply the number of corners times the number of courses. Basement windows are usually steel and are set in special, grooved window blocks. For each window, you will need two full and two half window blocks.

If the house is built on a hillside and there will be pressure against one or more walls, use at least 10-inch block on these walls and reinforce with wire made for this purpose.

Measure for the foundation walls the same as you did for the footings. Be very accurate because the top of the wall will determine the shape of the house. Level the same way as for the footings. (See *Footings*.)

Mortar can be mixed by hand or by a gasoline- or electric-powered mixer which can be rented. Your local dealer can estimate the quantities of cement and sand you will need. Ready mixed and sacked mortar is also available.

Mortar should be on the firm side and laid ⅜ inch thick. If the footings are not level, the bottom course of block will have to take up the error. Start at the highest corner and use a thick layer of mortar.

Laying blocks takes some skill. Using a trowel, lay two rows of mortar on the footings. Then stand a block on end and put mortar on it. Turn the block on its side and put it in place. If the mortar is too wet or too dry, it will probably fall off the block. After some practice, you will get it right. By the time you get to the top, you will have a presentable wall. After laying each block, tap it into place with the trowel handle until the mortar is about ⅜ inch thick. Then check for straightness and levelness. If your house has a basement and a garage, practice on the garage foundation, since it will not show.

Start all corners first, then lay out a string of blocks for each wall, correctly spaced to determine where the blocks will lie. Start the corners so that the top course on the end walls will have a joint in the center of the wall. You can have your concrete block dealer cut the blocks with a special saw where required. You can also rent a concrete block saw which is especially handy. Stretch a string out between the corner blocks to keep each wall straight. Use a straight 2 x 4 about 12 feet long and a level to keep the walls straight. On each course, level all four corners by stretching a line across from corner to corner and using a line-level at the exact center.

On the next-to-top course of basement walls, place steel windows in the grooves of the window block. Keep the locations within 8 inches of those shown on the plan, but locate them so that the half blocks will be on the top course.

Before laying the top course, or the course below the windows or beam pockets, lay a strip of metal lath over the block to support the mortar fill on the next course.

On the top course, a pocket must be built into each end wall to support the center beam. Cut the blocks to form the pocket. The top course is then filled with mortar and ½-inch by 8-inch bolts placed in the mortar at 8-foot intervals and at all corners. See the foundation drawings for the location of the bolts so that they will line up with the sill. Also provide ½-inch reinforcing rods protruding from the side of the block wherever concrete drives, walks or patios will adjoin the wall. (See *Concrete Flatwork*.)

If you prefer a poured concrete foundation, call a foundation contractor and have him form and pour the entire foundation. Be sure he follows the foundation plans. He may want to provide the footings also. The stepped top on foundations for brick veneer houses is usually omitted when poured concrete is used due to the high cost of forming the step. If you are going to have a septic tank, be sure to leave a hole in the foundation for the building drain. (See *Septic System*.) Once again, provide ½-inch reinforcing rods protruding from the foundation wherever concrete walks, drives or patios will adjoin the wall. (See *Concrete Flatwork*.)

Attach metal window wells to the basement walls with concrete nails before backfilling. The bottoms of window wells will be filled with gravel.

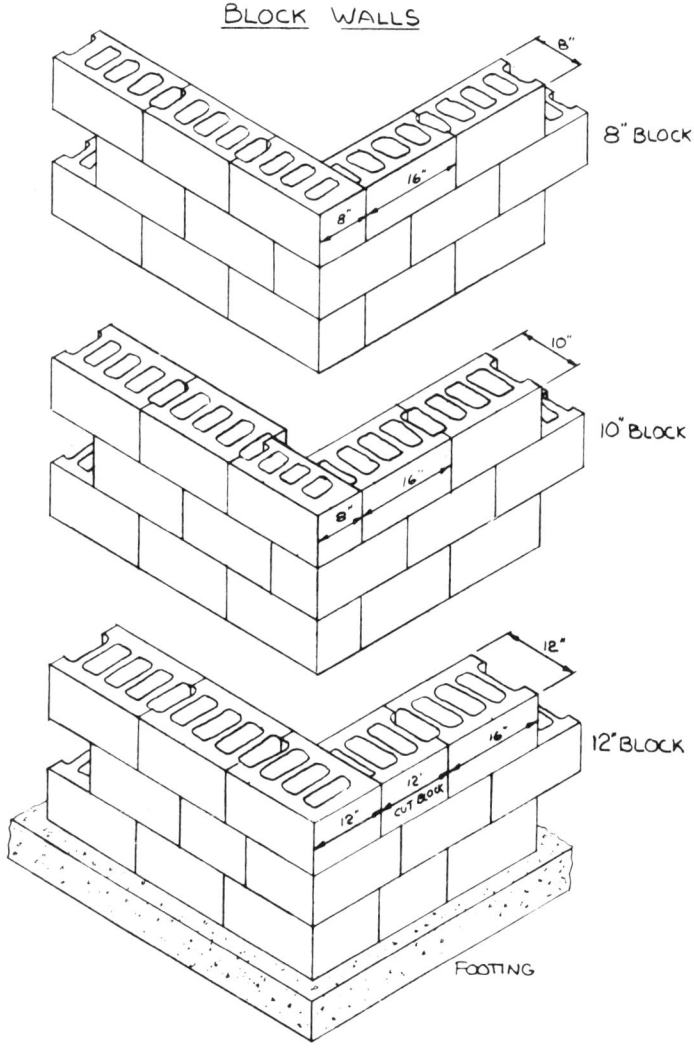

FOUNDATION WALL

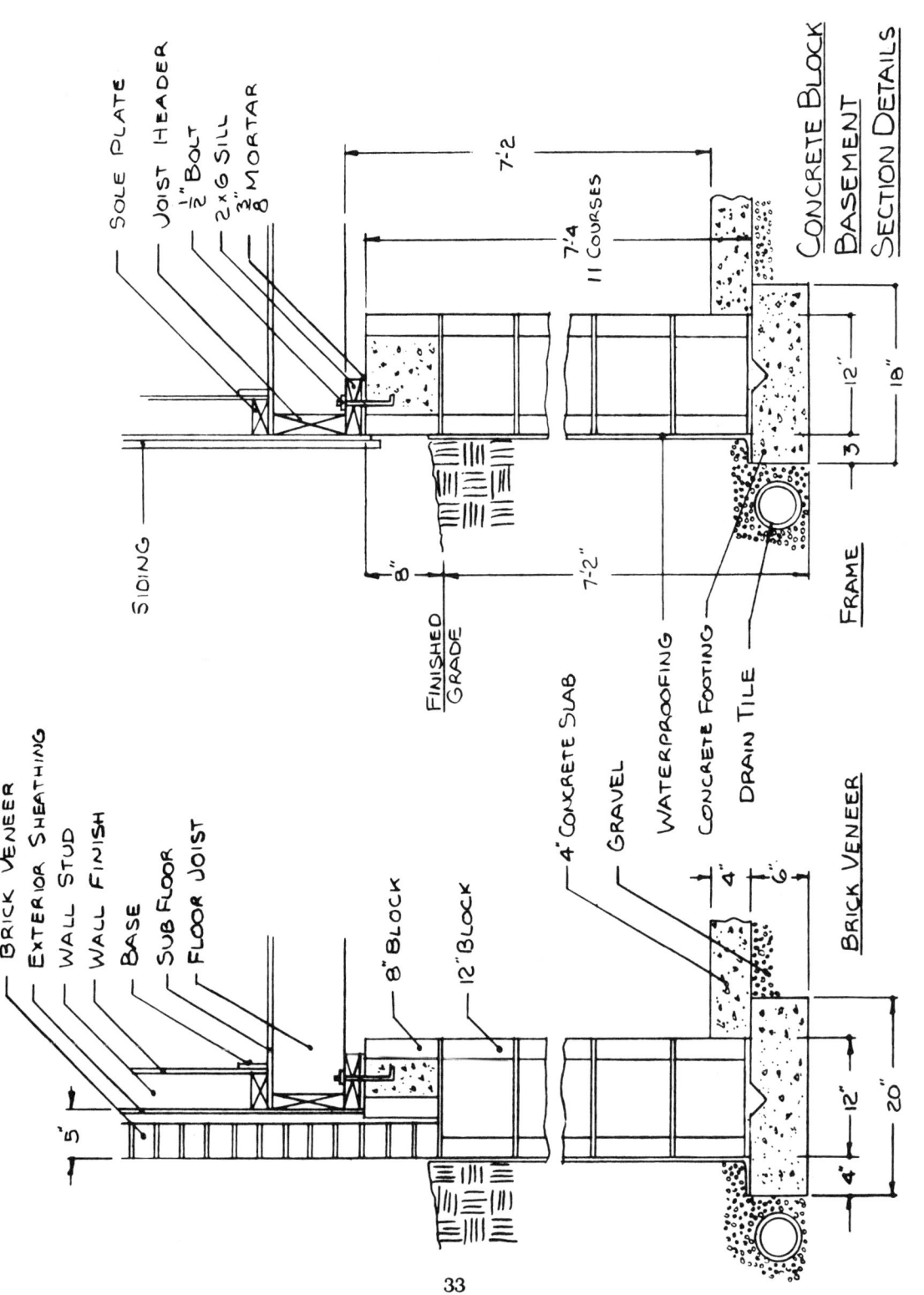

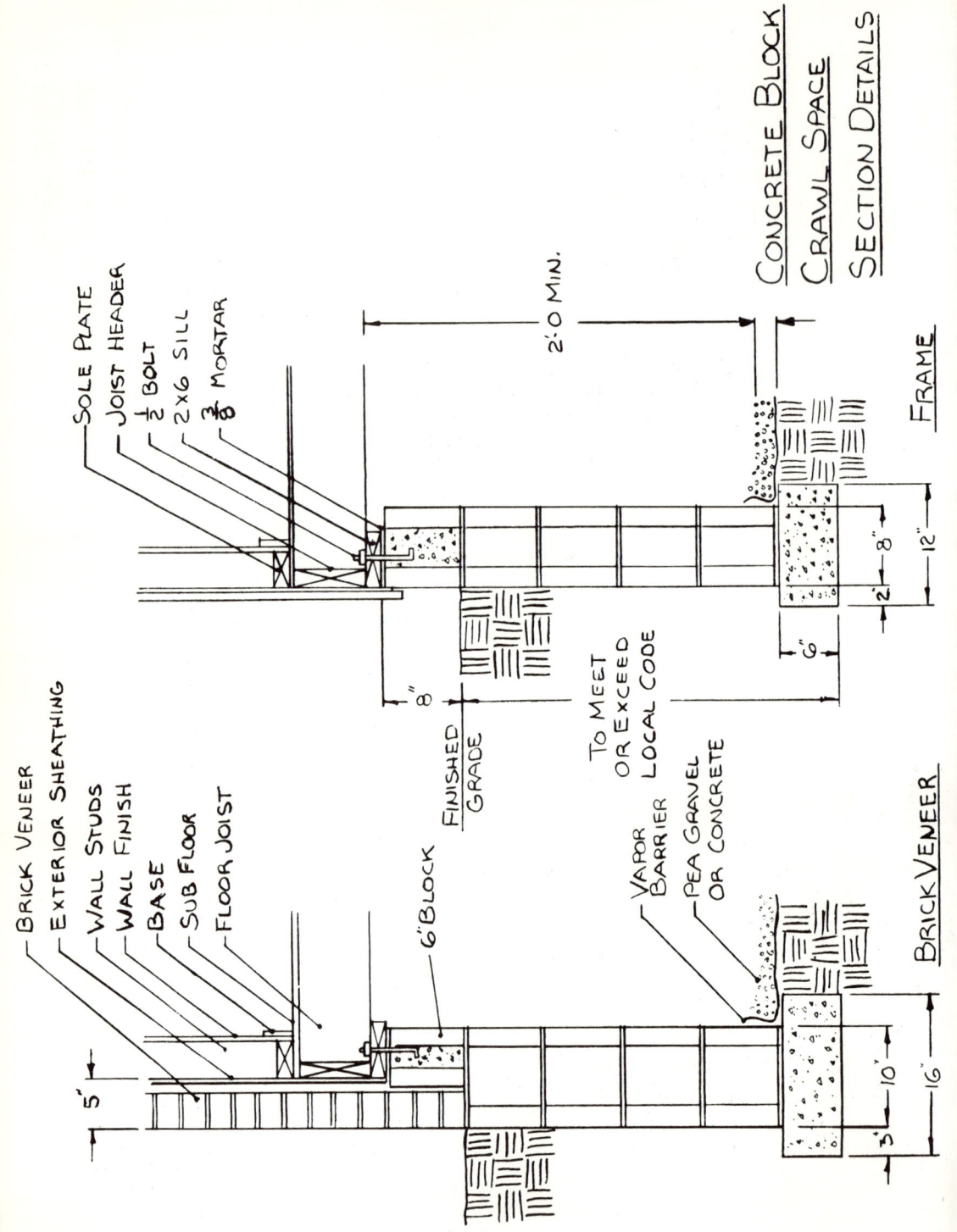

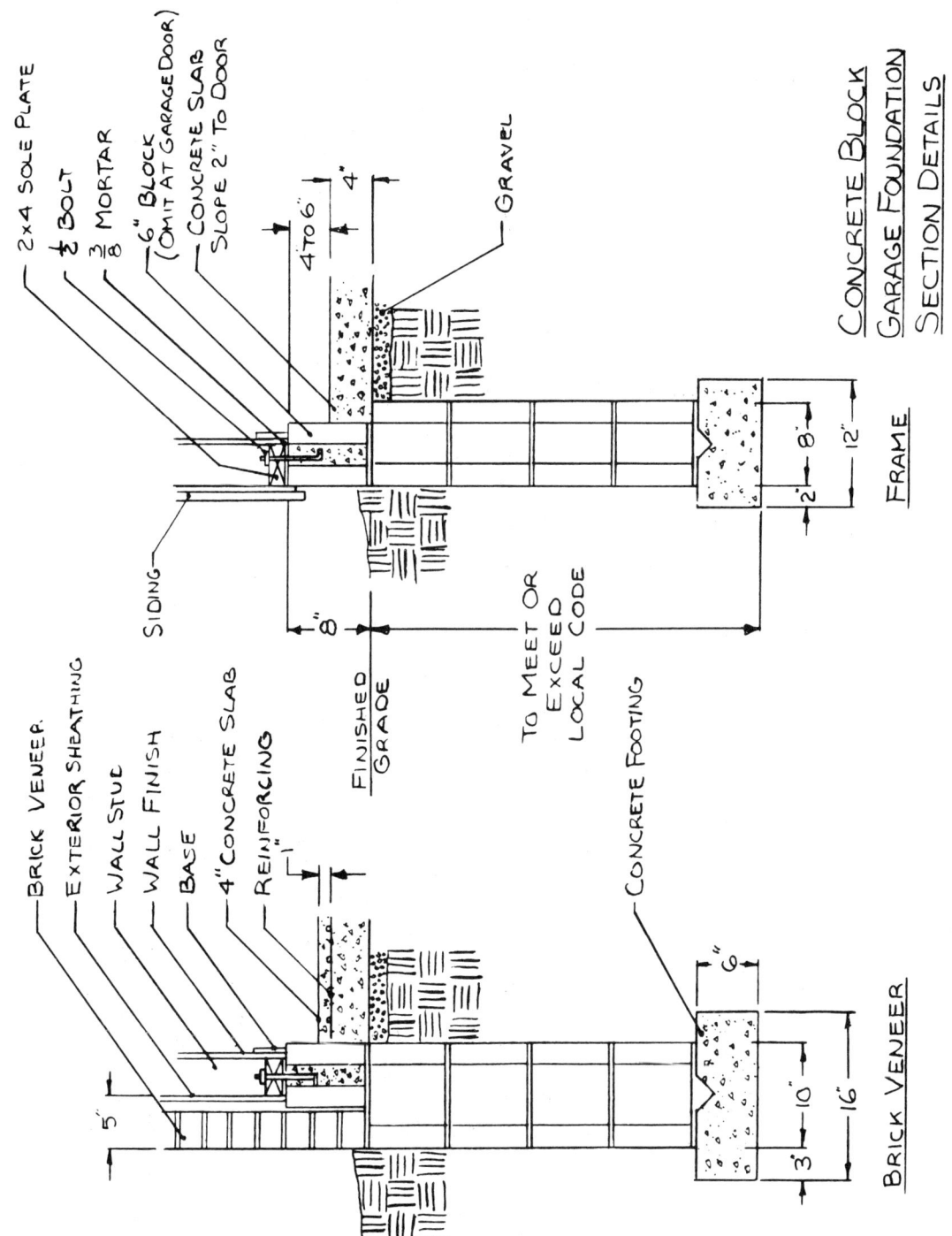

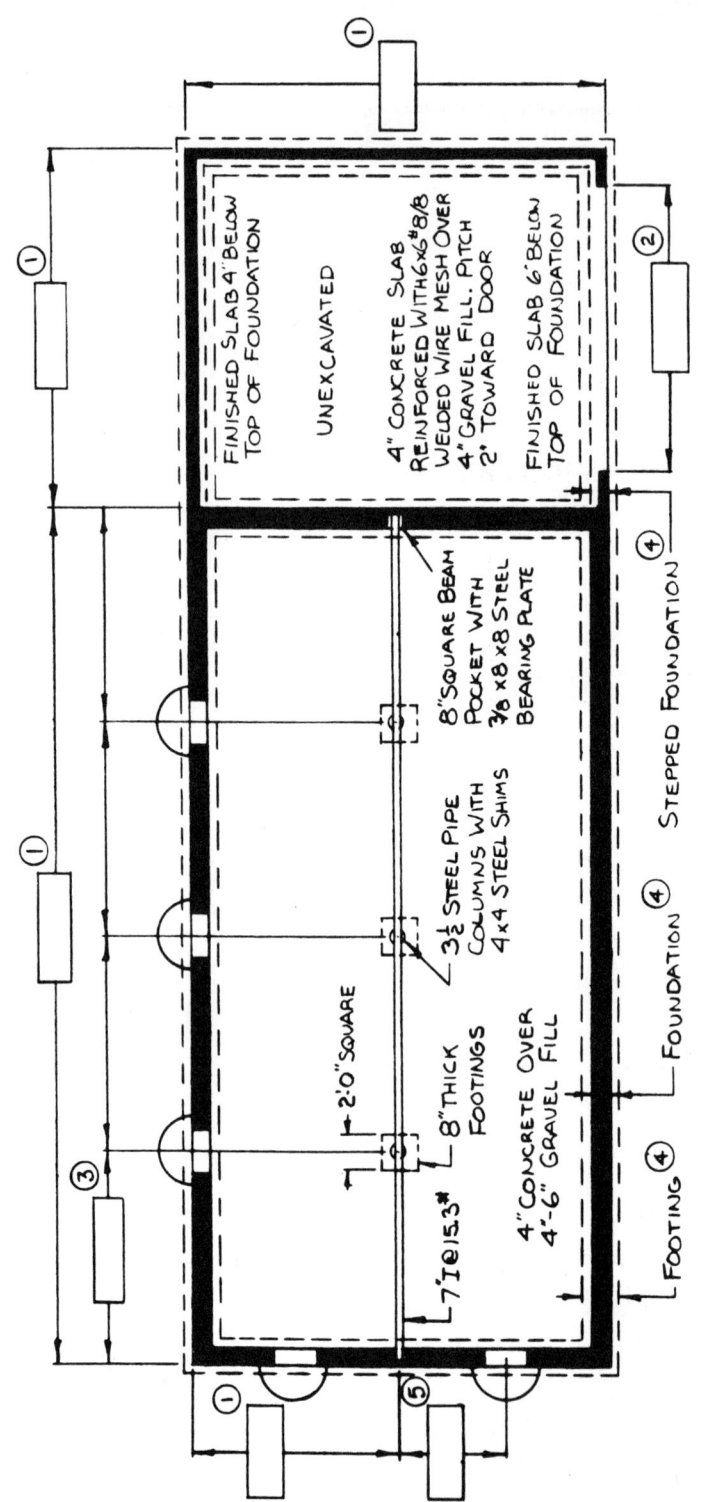

Foundation Plan
Basement & Garage

1. From Floor Plan. (Add 5" for Brick Walls)
2. From Floor Plan. Add 4' to Door Width
3. Width From Floor Plan – 20'-24' 25'-36' 37'-48' 49'-60'
   No. of Equal Spaces – 2   3   4   5
4. See Foundation Section Details' for Dimensions
5. From Floor Plan, Under House Windows

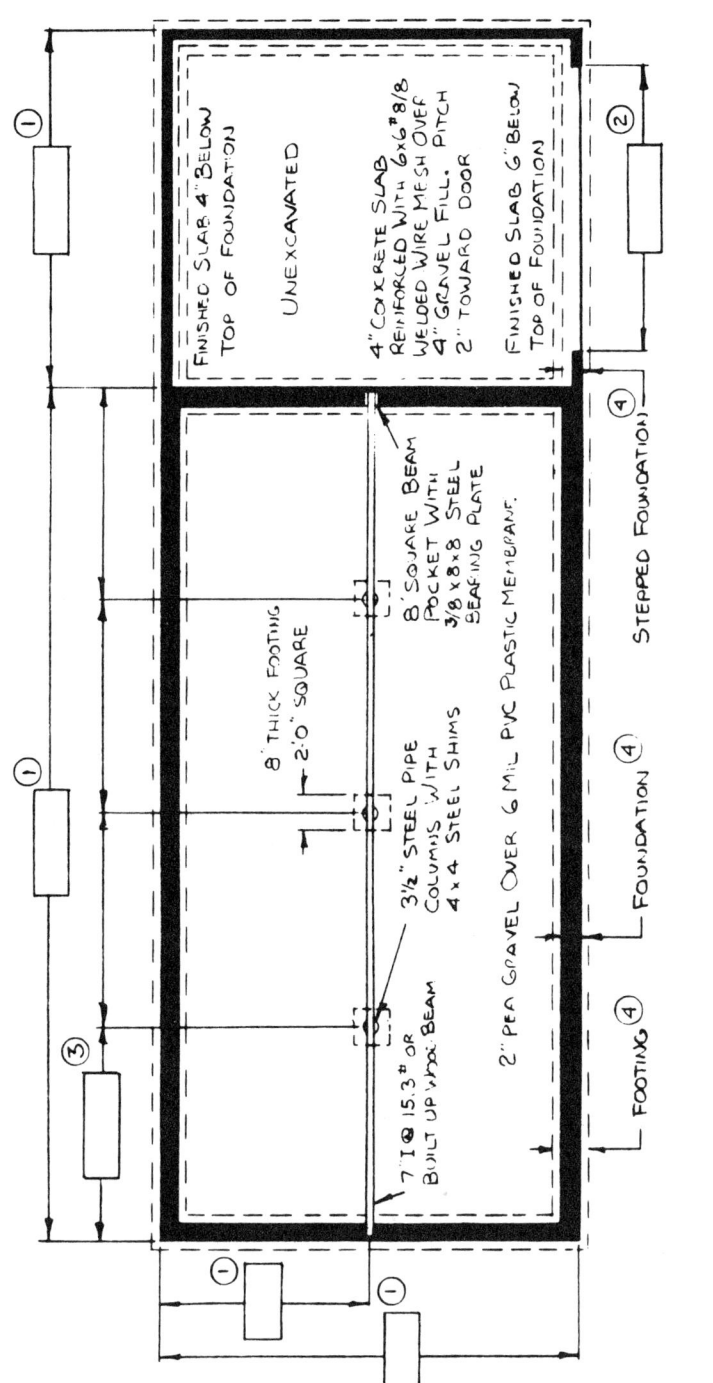

FOUNDATION PLAN
CRAWLSPACE & GARAGE

① FROM FLOOR PLAN. (ADD 5" FOR BRICK WALLS)
② FROM FLOOR PLAN, ADD 4" TO DOOR WIDTH
③ WIDTH FROM FLOOR PLAN - 20'-24' 25'-36' 37'-48' 49'-60'
   ROOF EQUAL SPACES -   2   3   4   5
④ SEE 'FOUNDATION SECTION DETAILS' FOR DIMENSIONS

12. Waterproofing the Foundation

The floor of the crawl space must be covered by a vapor barrier. Clear and level the ground and cover with a 6 ml. PVC (polyvinyl-chloride) plastic membrane. Purchase the widest roll available. If a seam is required, overlap at least 12 inches. Cover the vapor barrier with about 2 inches of pea gravel. Eight cubic yards will cover 1200 square feet.

If the house has a basement, a 4-inch perforated plastic or fiber drain pipe should be laid around the footing. Check with local contractors to see if it is required in your area—some soils drain well and a drain pipe is not needed. If it is required, put down about 2 inches of gravel all around the house next to the footing. Pitch the pipe to let water run off by gravity to a pit about 4 feet square and 4 feet deep filled with crushed rock. Cover the pipe with 8 inches of gravel.

If the house has a poured concrete basement, cover the outside of the foundation wall with a thick coat of bituminous dampproofing material, including the top of the footings. Coat up to the finished grade level.

If the house has a cement block basement, the walls should be plastered with at least ⅜ inch of Portland cement up to the finished grade. Then apply a coat of bituminous dampproofing material, including the top of the footings. An alternate method is to coat the cement block walls with dampproofing material, and cover with 6 ml. PVC plastic membrane while it is still sticky. Then cover the membrane with another coat of dampproofing.

After the wall has had about a week to dry, the earth can be backfilled around the foundation. This will make it easier to work around the house while framing the floor.

BASEMENT

CRAWL SPACE

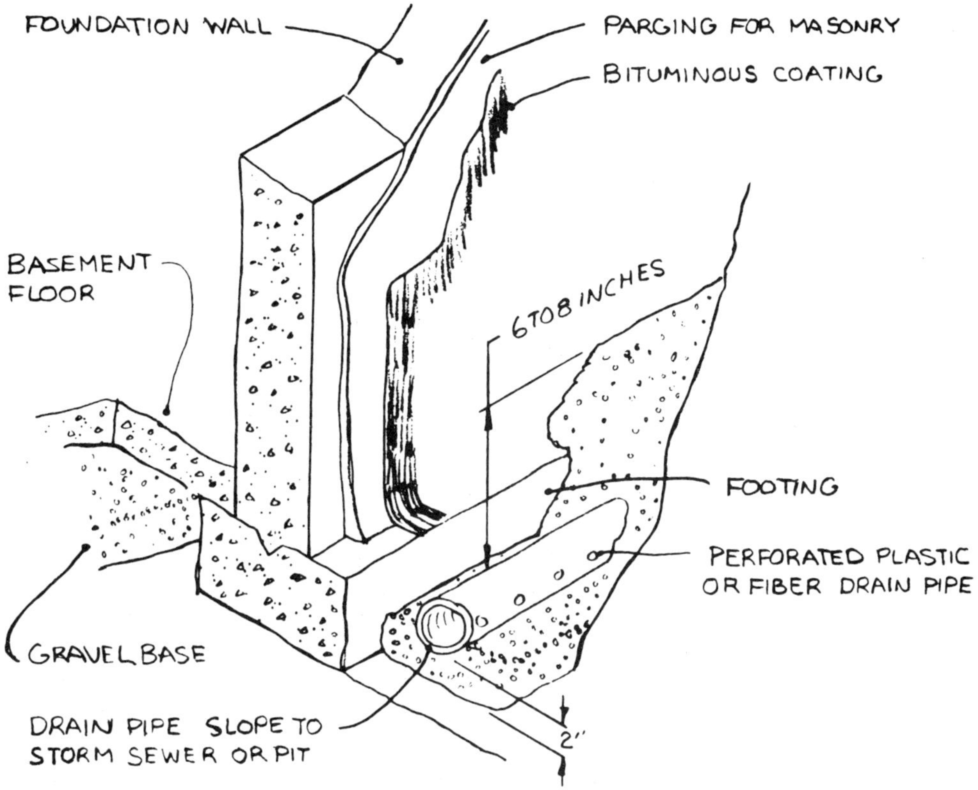

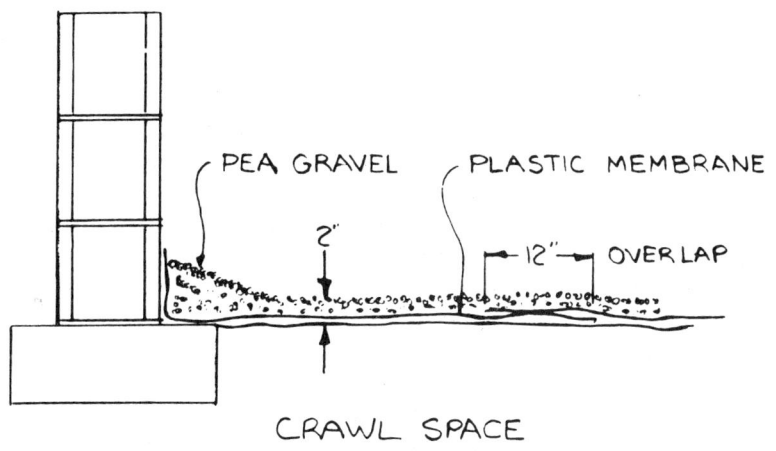

WATERPROOFING THE FOUNDATION

13. Slab Floors

A slab floor is the most economical floor, and if installed correctly is completely satisfactory. Unless you are a skilled concrete finisher, make an appointment with an expert concrete finisher to finish the slab. The sewer pipes must be installed under the slab. (See *Plumbing and Drainage*.) Also install the furnace pipes if they are to be in the floor. (See *Heating: Perimeter for Slabs*.)

Remove any vegetation, level the earth in the slab area, and cover with a 6 ml. PVC plastic membrane. Overlap the joints at least 12 inches. Place the perimeter insulation as shown in the foundation drawings. The thickness of insulation will vary with the climate and type of insulation used—your local heating dealer will know what you need. Cover the entire area with gravel to within 4 inches of the top of the foundation, except for a 24-inch-wide strip down the center of the house, under the center partition. This is to form a concrete beam to support the center partition. Cover the entire area with a 6- x 6-inch reinforcing mesh. Overlap joints 6 inches. The cement contractor may want to do this himself.

Have the slab poured and finished flush with the top of the foundation. This finish must be as smooth and level as possible.

An excellent floor can be made without the help of a concrete finisher if desired. Two methods are described below. Be sure you have the approval of the building inspector since these methods are not generally covered by building codes.

Follow the directions for *Slab Floors*, except pour the floor yourself using ready mix cement if available. Place a 2 x 8 down the center of the floor to divide the floor into two 12-foot widths. This board must be straight and level with the top block. Smooth the concrete flat and level with the top of the footings. This is a two-man job. Lift the wire mesh as you go along. It should rest

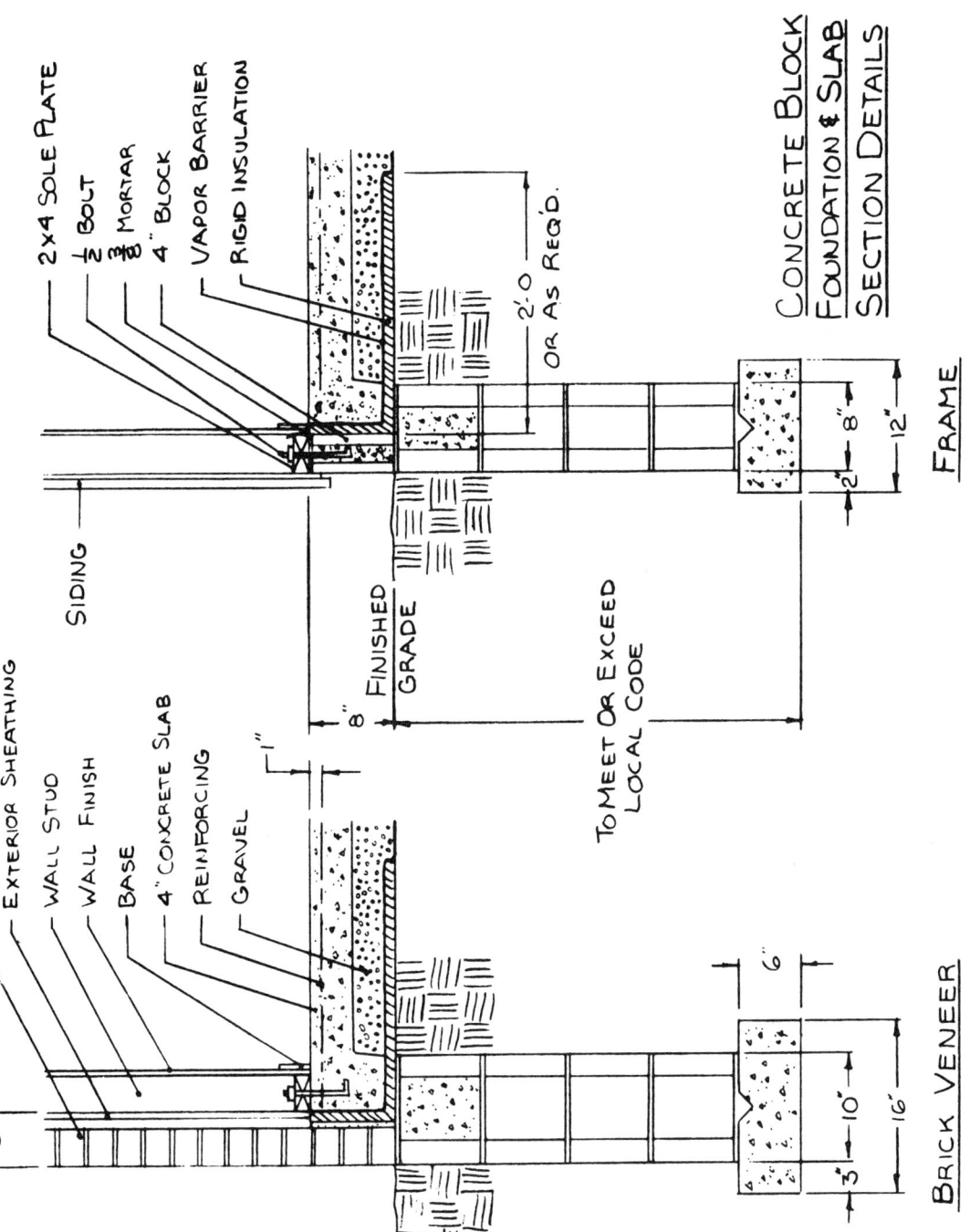

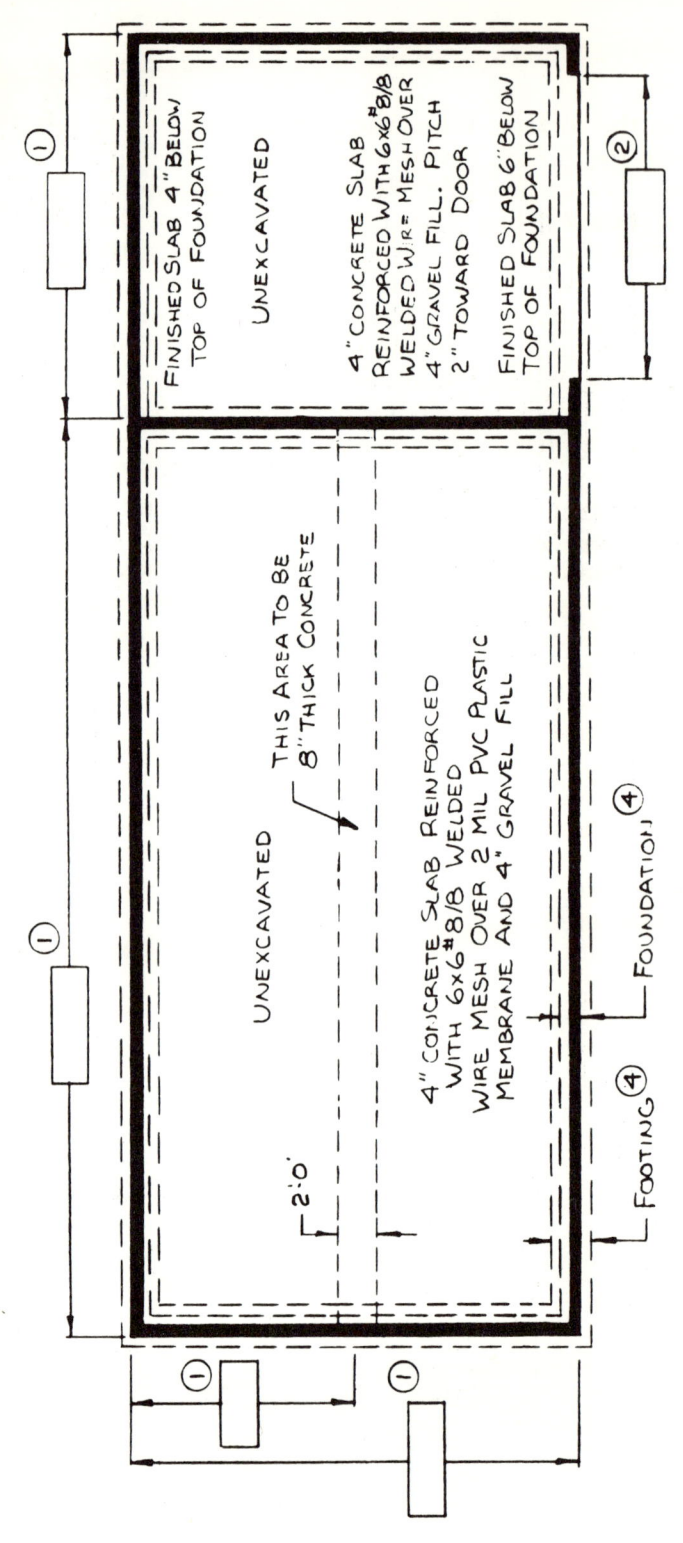

about 1 inch below the surface. When the first half is firm enough to walk on, remove the 8-inch board and pour the second half. Keep the floor as level as possible. Remember, the floor of the house is very important. (See *Concrete Flatwork* for further details.) When the concrete has set, cover the entire area with 2 x 2 lumber in the same pattern as shown on the floor framing drawings. This can be nailed to the slab with special masonry nails and a driver sold by hardware or lumber dealers. Apply the subfloor. (See *Floor Framing and Subflooring*.)

Another type of wood floor on a slab is even faster and easier to build, but slightly more costly. This is a conventional floor built over a shallow, concrete covered space. Follow the directions for *Slab Floors*. After the building drain and water are in, put down the insulation and plastic vapor barrier and then pour a thin 2-inch covering of concrete. Smooth this out as well as you can (have help to do it). No reinforcing is required. Frame the area with a conventional wood sill and floor joists (see *Floor Framing and Subflooring*). After the joists are all in place, sheet metal heating ducts for forced warm air heat can be installed. (See *Forced Warm Air Heating*.) The subflooring can be installed as soon as the heating is complete.

14. Pier Foundation

In many parts of the country, particularly in the south where the winters are mild, it is customary to build foundations of wood posts or concrete block piers. This type of foundation is also popular for vacation homes, because it can be built much faster than a slab or crawl space, and at a lower cost.

The site is prepared by digging a series of holes for the posts or piers. (See *Preparation of Site* and *Footings*.) The locations are shown on the foundation plan. For wood posts, pour 20-inch-square footings, 8 inches thick. For concrete piers, pour the footings as shown on the foundation section details. No forms are required, but the tops of all the footings must be level.

Wood posts should be either 8 inches in diameter if round or 6 x 6 inches if square, and they must be treated to prevent rotting. Round posts should be squared at the top to 6 x 6 inches to the depth of the edge beam, usually 11¼ inches. Place the posts in the holes, and nail a temporary 2 x 4 along the sides of the posts to align them. Check and recheck the dimensions against the foundation plan and be sure the diagonal measurements are of equal length. When this is done carefully, replace the dirt around the posts. Break the dirt into fine pieces and tamp into the holes evenly so as not to disturb the posts.

Piers make a more permanent-looking foundation than wood posts, and are recommended. The piers on the ends of the house are 16 x 16 concrete block, and the others are 8 x 16. Fill the top three courses with concrete, and put a ½-inch anchor bolt in each pier.

The edge beam is installed next. This double 2 x 12 beam goes all around the house, and down the center. Nail the two pieces together with 10d nails 16 inches apart in two rows. Make sure the beam is level, straight and square, and made to the correct

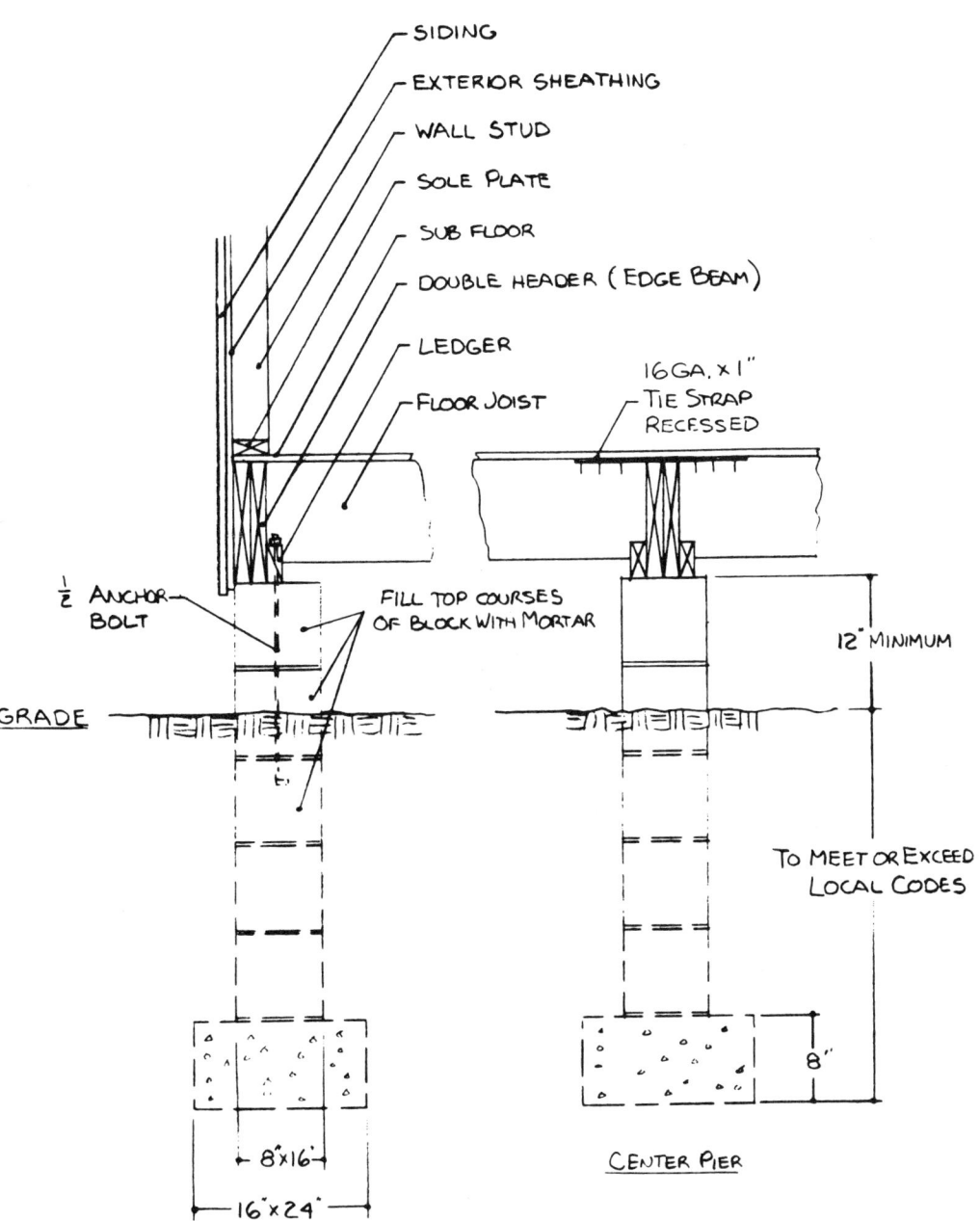

PIER & EDGE BEAM FOUNDATION
SECTION DETAIL

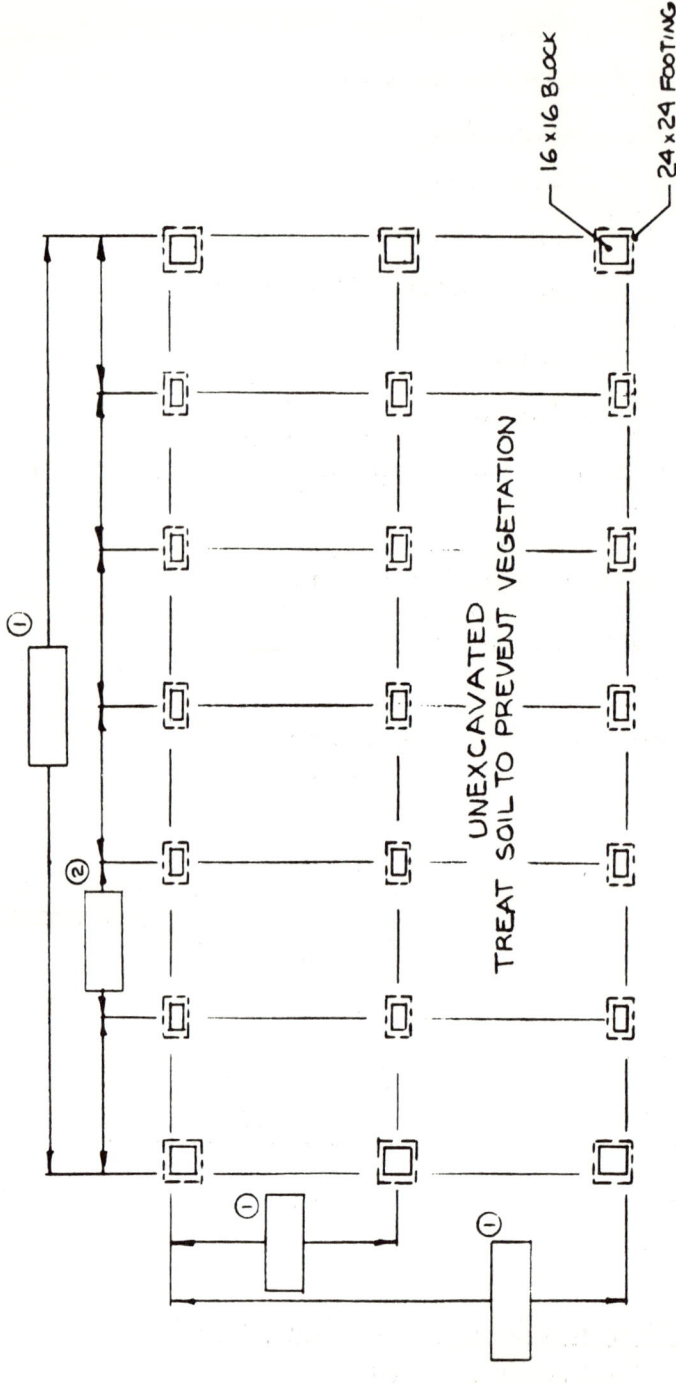

dimensions shown on the floor framing plan. Next, nail the 2 x 4 ledger in place. For post foundations, toenail the ledger to the top of the posts. For pier foundations, drill holes in the ledger and bolt it down with the ½-inch anchor bolts in the piers. Nail the ledger to the edge beam with 16d nails staggered 16 inches apart.

Next, cut the floor joists so they fit snugly between the edge and center beams. Notch each joist so that it rests on the ledger and is flush with the tops of the beams. Nail in place using 30d nails through the double edge beam. Use a piece of 16 gauge x 1 inch steel strap to join the joists together across the center beam. Recess this flush, so the floor sheathing will lie flat. Next lay the subflooring. (See *Floor Framing and Subflooring.*)

The pier or post foundation has a 12-inch or larger open space around the bottom which is exposed to the weather. This may be left open or covered with a skirt board made of ½-inch or larger exterior plywood. The skirt board must still provide an inch or two of ventilation space above the ground. Usually the space is left open, and if insulation is required, it is a rigid type which is nailed between the joists, against the floor sheathing. The water and soil pipes from the house must also be insulated in freezing climates. This can be done by wrapping them with a waterproof insulation down to about two feet below the grade level.

15. Floor Framing and Subflooring

Spread a layer of mortar on top of the foundation. Place the termite shield, which comes in rolls, on top of the mortar. Cut holes so it will go over the anchor bolts. Termite shield should be continuous with breaks only at the corners. If termites are not a problem in your area, and if the code permits, you may omit the shield.

If a termite shield is not required, the sill may be put down without mortar. A 1½-inch layer of fiberglass insulation should be placed under the sill to fill any gap. This will keep wind from blowing into the basement or crawl space.

Lay the sill on top of the foundation and mark for the anchor bolt holes. Drill the holes. See the foundation drawings for the size of the sill. The sill must be level; if it is not, it should be shimmed level. Nuts and washers are placed on the bolts to hold the sill down.

Buy 3-inch steel pipe posts of the correct length to support the center beam. They will rest on the concrete piers in the basement or crawl space. See the foundation plan. Posts are made for either steel or wood beams. Get the proper type. Also get some 4- x 4-inch steel shims to use to support the posts. The post should be about 6 inches shorter than the distance from the top of the piers to the top of the foundation.

If the house has a basement, **buy a** steel beam (7" I @ 15.3#) of the correct length to reach from pocket to pocket. Measure carefully. Have the beam delivered and set into place, as it is too heavy to be lifted. Use an 8- x 8-inch steel bearing plate, ⅜ inch thick, to support the beam in each pocket.

A wood beam can be used for crawl spaces. Build up the beam with three thicknesses of 2 x 12 lumber at least 12 feet long. Nail from each side with two 20d nails every 32 inches. Build the beam in place by setting a 2 x 12 on top of a post and holding it up with

FLOOR FRAMING AND SUBFLOORING 49

temporary braces. Then nail the other pieces to it. The joints should be staggered so that no two joints are closer than 4 feet apart.

Shim the beam until it is perfectly level and even with the top of the sill. It must also be straight down the center of the house and an equal distance from both walls.

Use 2 x 8 floor joists for spans of 10 to 12 feet. Use 2 x 10 floor joists for spans of 14 to 16 feet. Lay all the floor joists across the sills and beam. Nail the headers in place all around the sill. Toenail a 10d nail every 16 inches into the sill. Stretch a steel tape along the header and mark 16-inch spaces as shown on the floor framing plan. Nail the floor joists in place using two 20d nails through the header. Where the joists meet over the center beam, nail them together with two 10d nails. If a steel beam is used, the joists will have to be lined up when the subfloor is applied.

Frame the headers and double joists around basement stairways as shown on the floor framing plan. Nail with three 20d nails through the headers into the joists and then nail the second header to the first with 16d nails staggered and spaced 6 inches apart. Nail the double joists together with 16d nails staggered 6 inches apart.

Cut the bridging as shown in the floor framing plan and nail at the top ends only, using two 6d nails. The bottom end should be nailed after the subfloor is in place.

Lay ⅝-inch exterior plywood subflooring over the floor joists as shown on the floor framing plan and nail with 8d nails every 6 inches all around the edge and every 12 inches on the center joists.

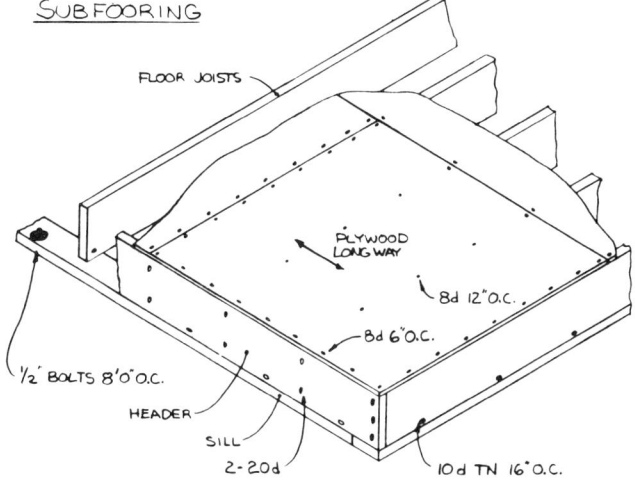

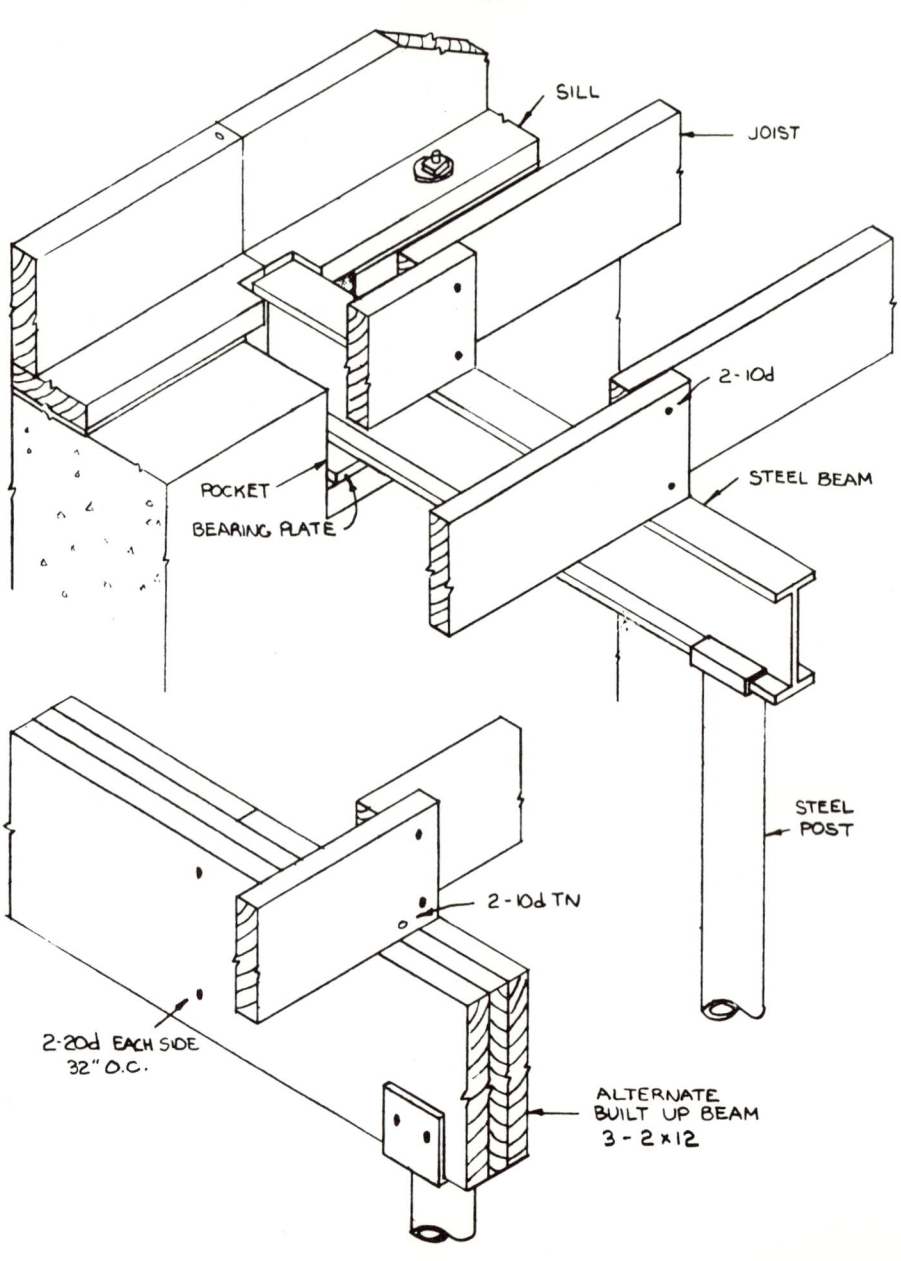

FLOOR OPENING

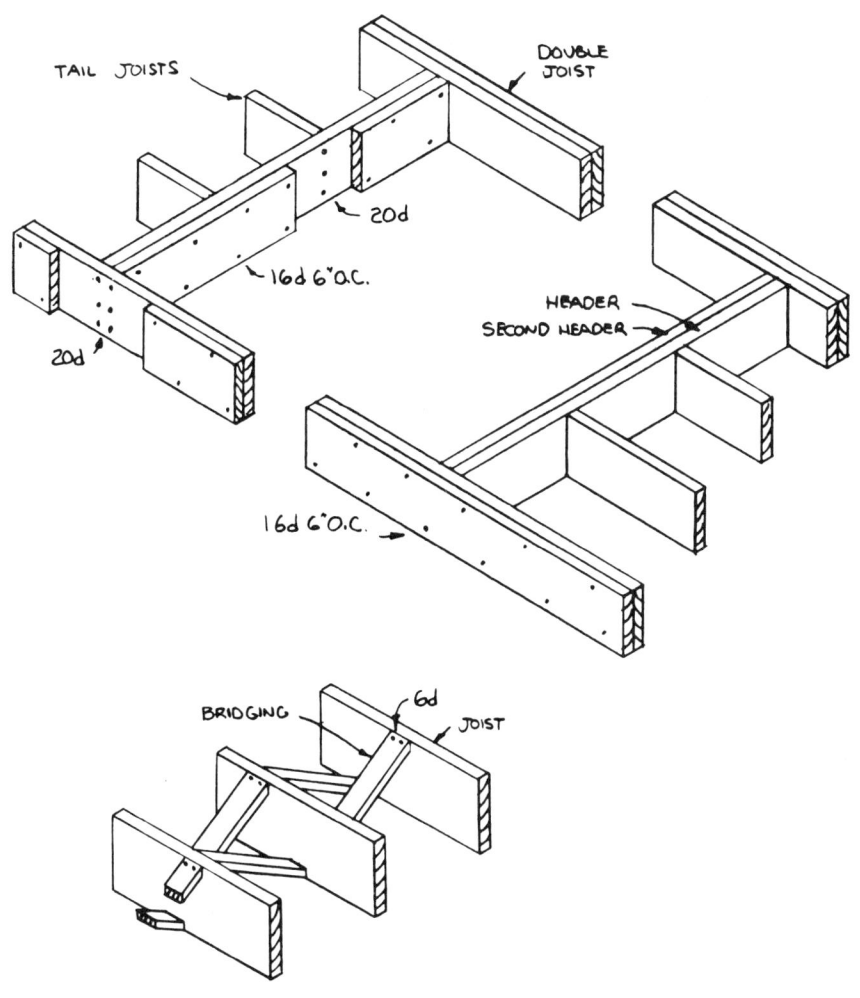

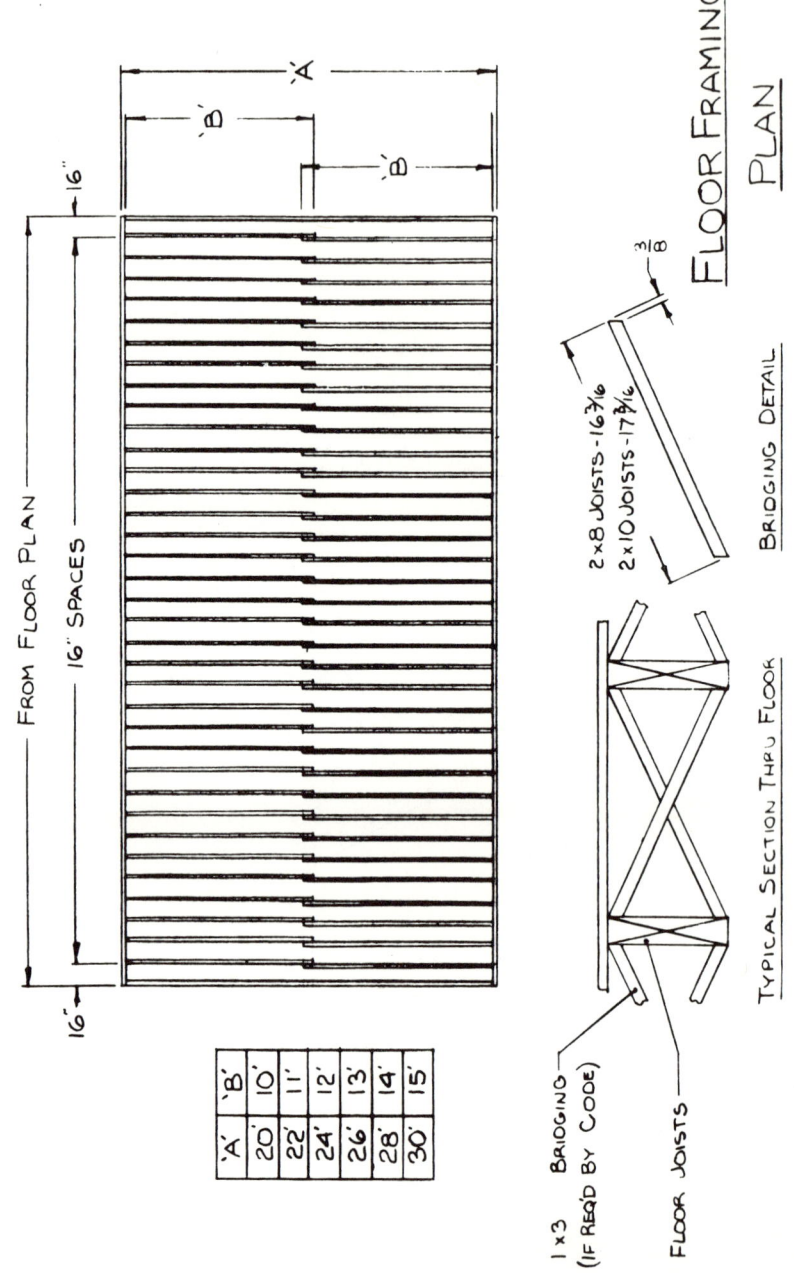

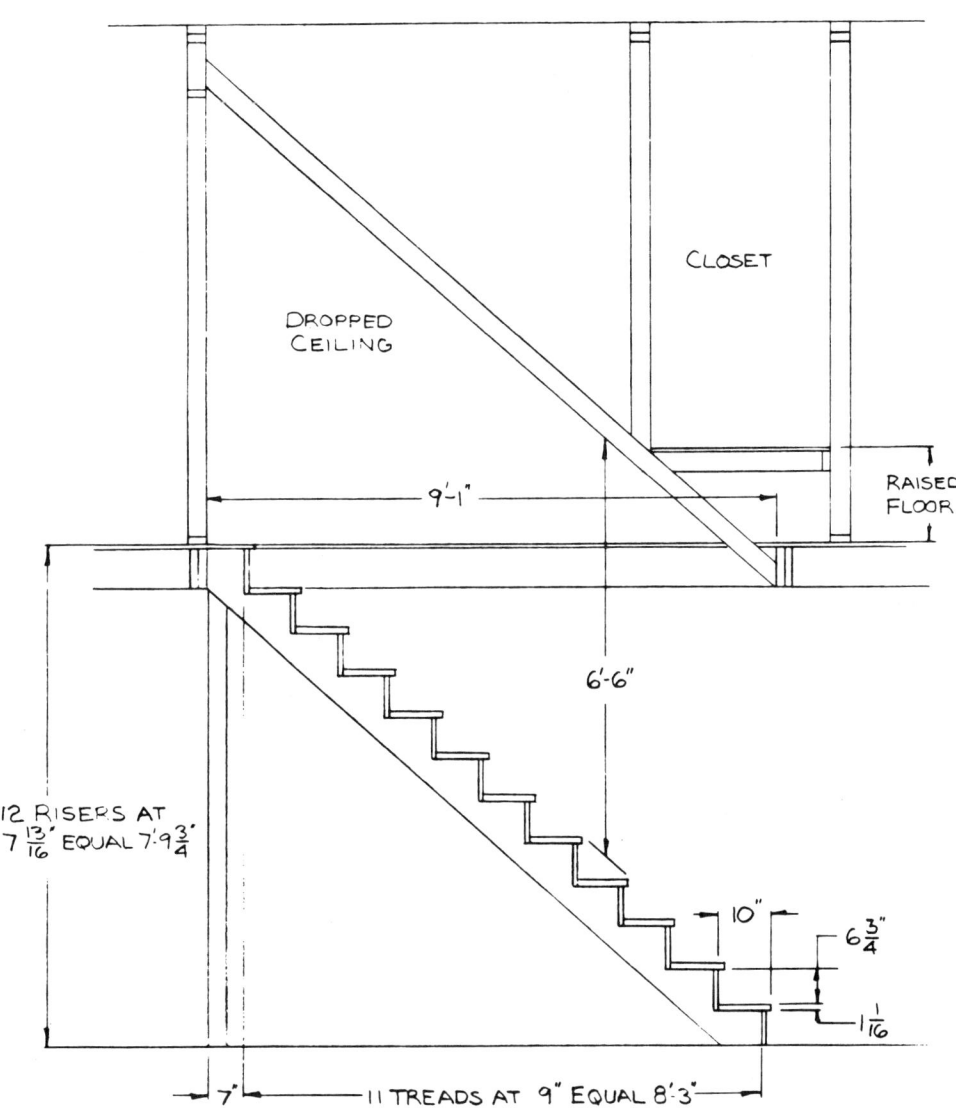

STAIRS WITH CLOSET ABOVE

16. Wall Framing and Sheathing

The exterior wall framing is 2 x 4 material. The walls are laid out and assembled on the floor. The studs are usually available ready cut to size from lumber dealers. The correct length for most plans is 7 feet 9 inches. This will make 8-foot ceilings. Cut the pieces to the sizes shown on the external wall framing details.

Build the front and rear walls first; they extend the full length of the house. For example, if the floor plan shows that the length of the house is 40 feet 0 inches, the wall will also be 40 feet 0 inches. Lay the sole plate and lower top plate side by side and stretch a steel tape along them; mark off 16-inch spaces as shown on the exterior wall plan. Use a carpenter's square to mark across both pieces. Next measure and mark the door and window spacing. These dimensions are shown on the floor plan and are measured to the center of the door or window. These marks should be marked "CL" for center line, to avoid confusing them with the stud center lines.

Next, lay out the rough opening width. For windows, this width should be obtained from the window manufacturer or window dealer. For doors, the rough opening width is 3½ inches wider than the door. Measure half the rough opening width from each side of the center line. Locate all intersecting walls on the floor plan and measure and mark their location on the sole and top plates. Mark the 3½-inch space where an intersecting wall will go with a "W" for wall.

Separate the sole plate and lower top plate and lay the studs in between. Nail the studs in place with two 16d nails through each plate. Omit the studs in the areas where the doors, windows, and the intersecting walls go. The studs which frame the rough openings are cut 6 feet 9¾ inches long to support the 2 x 12 headers which measure 1½ inches by 11¼ inches. Cut the 2 x 12 headers 3 inches longer than the rough opening dimension and nail them together with a ½-inch plywood filler between them for a spacer. Use 10d nails staggered 16 inches apart. Cut the sill piece for window to the rough opening width and nail between the studs with two 10d nails in each end. Next, nail two full-length studs to the 6 foot 9¾ inch studs with 10d nails staggered 16 inches apart. Nail three 10d nails into the end of each 2 x 12 header. Fill in the area under the window sill with short pieces of 2 x 4 located on the 16-inch space. The completed wall must have a stud on every 16-inch space. In the area where the intersecting walls go, nail two studs with three 2 x 4 x 12-inch spacers lying flat in between. Place these spacers toward the side where the adjoining wall will be nailed. Nail each spacer in place with two 10d nails through each stud. Where the front and side walls connect, the 2 x 4 x 12 spacers are placed on edge to form the corner.

Nail the upper top plate to the lower top plate. At every wall intersection, leave a 3½-inch space in the upper top plate to tie in the intersecting wall. Also, leave a 3½-inch extension on each end. The front and back walls will have the top plates cut back 3½ inches to accept the extension.

After the front and rear walls are done, build the end walls. The end walls are shortened 3½ inches on each end to fit up to the front and rear walls. The 16-inch centers are started out by placing the first stud 12½ inches from the end. This 12½ inches plus the 3½-inch thickness of the wall will make an even 16-inch space to the first stud.

When all four exterior walls are completed, square them up by making the diagonal measurements match. Then nail temporary 12-foot 2 x 4 braces diagonally from top to sole plate. Nail some short lengths of 2 x 4 to the floor at right angles to the long walls about 8 feet from each wall. Space them about 12 feet apart. Get some help to set the walls into place and nail temporary 12-foot 2 x 4 braces from the top of the wall studs to the floor. Nail the sole plate to the floor header with 16d nails 16 inches apart. Nail the overlapping top plates with two 16d nails. Stretch a line along the top plate of each wall and move the braces in or out until the wall is straight. Add more braces if required to hold the walls straight.

When all exterior walls are up, apply the exterior sheathing. If the wall is to be brick veneer, provide a 30-pound felt continuous flashing over the header and sill and let this overlap the top of the foundation to within 1 inch of the edge. The exterior sheathing will go over this. A 25/32-inch thick insulation board can be used in place of the plywood sheathing to provide better insulation for brick walls. (See *Brick*.) If horizontal siding is to be used, ⅜-inch plywood is acceptable, but you may want to use ½-inch plywood to provide a firmer nailing surface. Apply according to the wall sheathing drawings. The sheathing should come down flush with the top of the foundation. Nail with 6d nails spaced 6 inches apart around the edges and spaced 12 inches apart in the center. This will tie walls and floor framing to the sill and provide an extremely stiff structure.

The interior partitions are built in the same manner as the exterior walls with a few exceptions. The 2 x 12 headers over the doors are not required, so 7¾-inch studs and 2 x 4 headers are used. The 2 x 4 headers also have a ½-inch plywood spacer. The interior door openings are not usually located on the plan, because they are located next to a wall and are self-locating. Bifold closet doors do not need a frame. The rough opening width is 1 inch wider than the door width. The partition supporting the ceiling joists is a load-bearing partition, so any doors wider than 3 feet must have 2 x 6 headers. When building the partitions, don't use any more studs than required. The partitions do not require any extra strength, and extra studs will make wiring difficult. When all the walls and partitions are in place, the sole plate in the door openings can be cut away.

As each partition is assembled and put into place, measure the location carefully from one of the exterior walls and be sure that the sole plate is straight. If any errors accumulate, they should be taken up by altering the closet sizes. Cross bracing will have to be removed as interior walls are installed, so interior walls should have temporary diagonal braces nailed to them to maintain squareness. Occasionally recheck the exterior walls at the top for straightness.

Garage wall framing is the same as the exterior house framing, except the studs are longer. If the floor joists are 2 x 8's, the garage studs are 8 feet 6½ inches long. If the joists are 2 x 10's, the garage studs are 8 feet 8½ inches long.

The garage door requires a special header. It is built of 2 x 3's with a 24-inch-wide ½-inch plywood web on each side. Glue the header together with waterproof resin glue and nail with 5d annular nails 6 inches apart on centers. The header extends completely across the front wall of the garage to give more bearing area on the wall studs. This also allows alternate door sizes to be used, such as two 8-foot doors in place of a 16-foot door.

The garage has no center load-bearing partition to support the roof, so a plywood beam is used to support it. (See *Roof Framing and Sheathing*.)

CORNER CONSTRUCTION

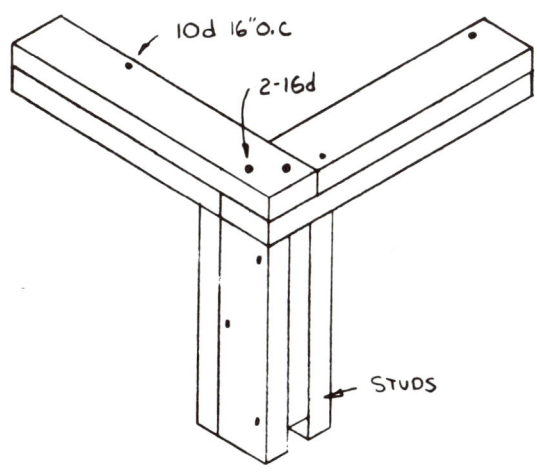

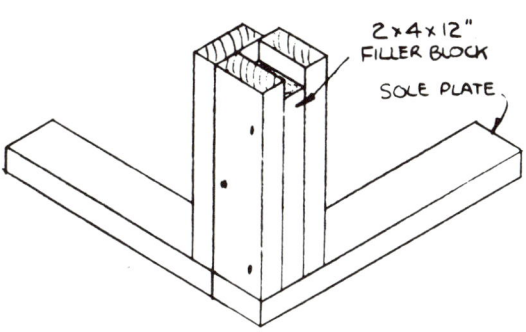

WALL INTERSECTION

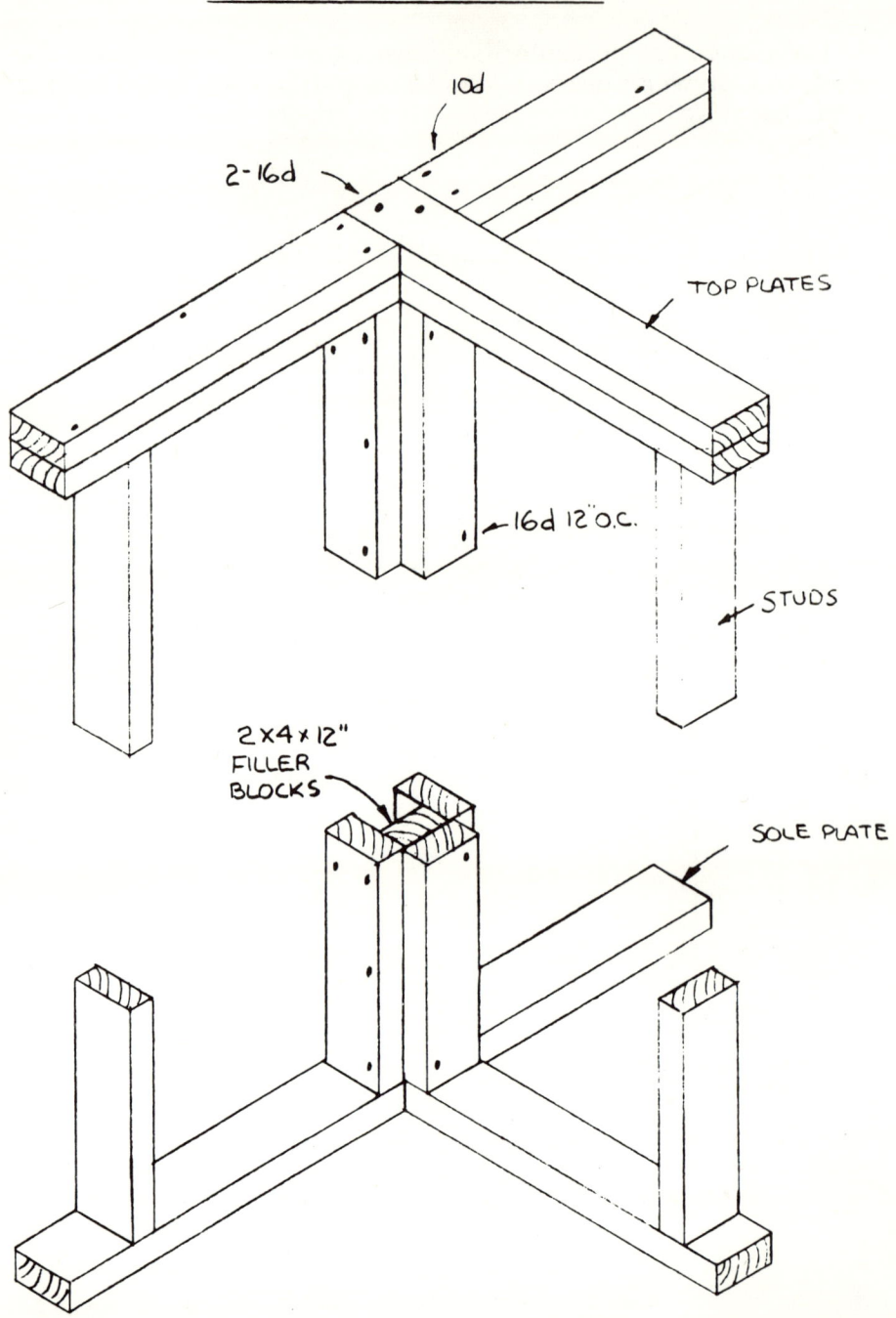

WINDOW FRAMING

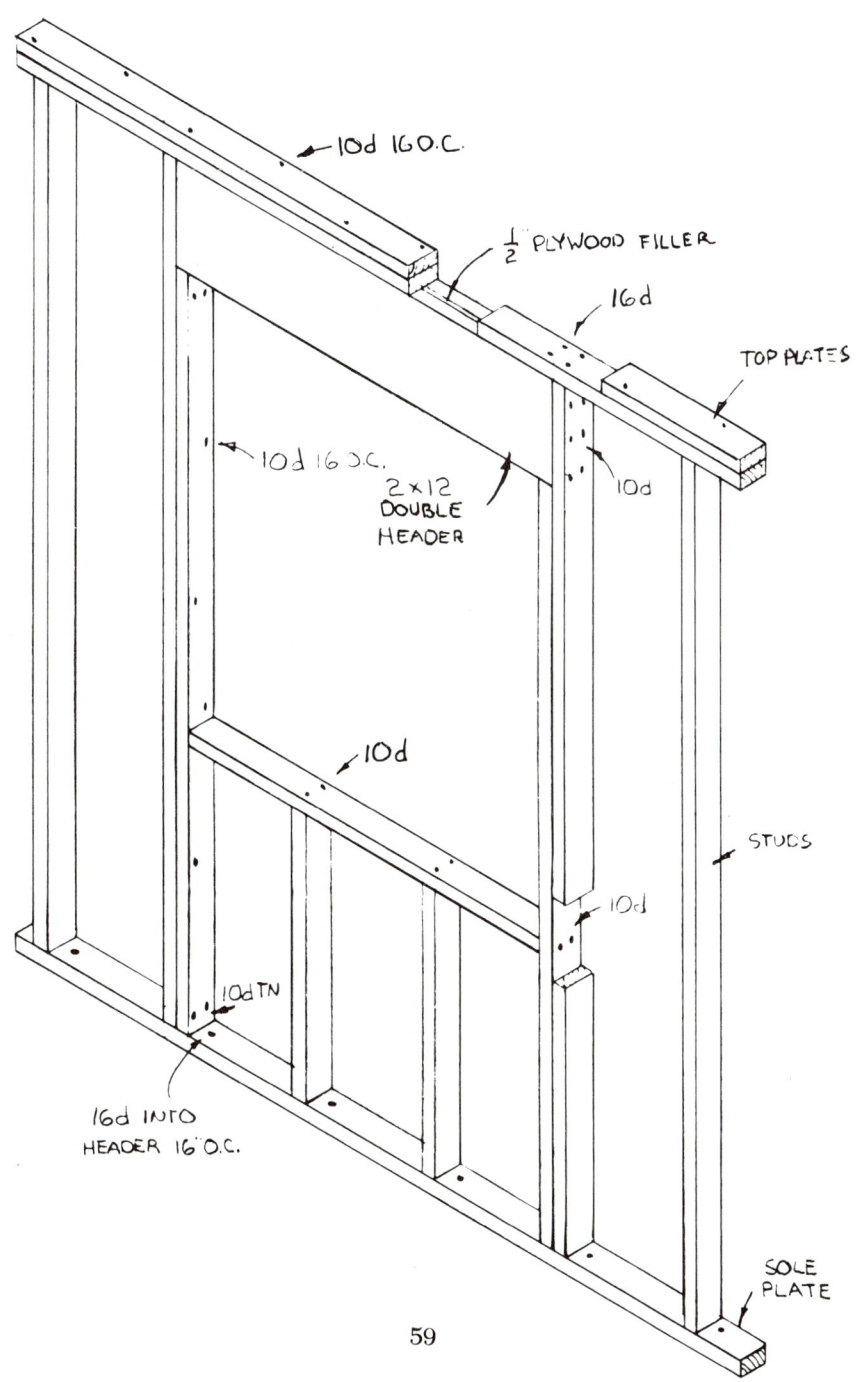

INTERIOR DOOR FRAMING

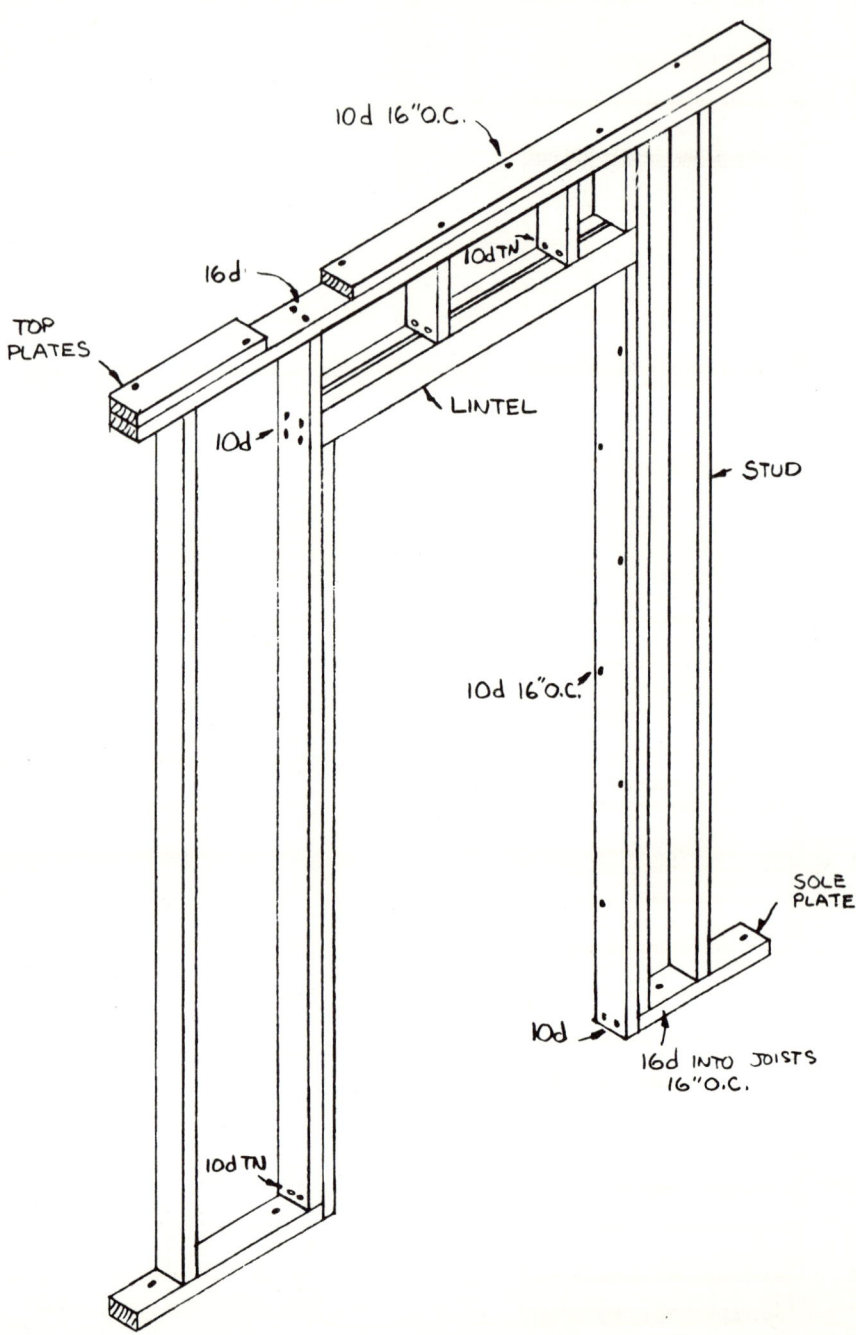

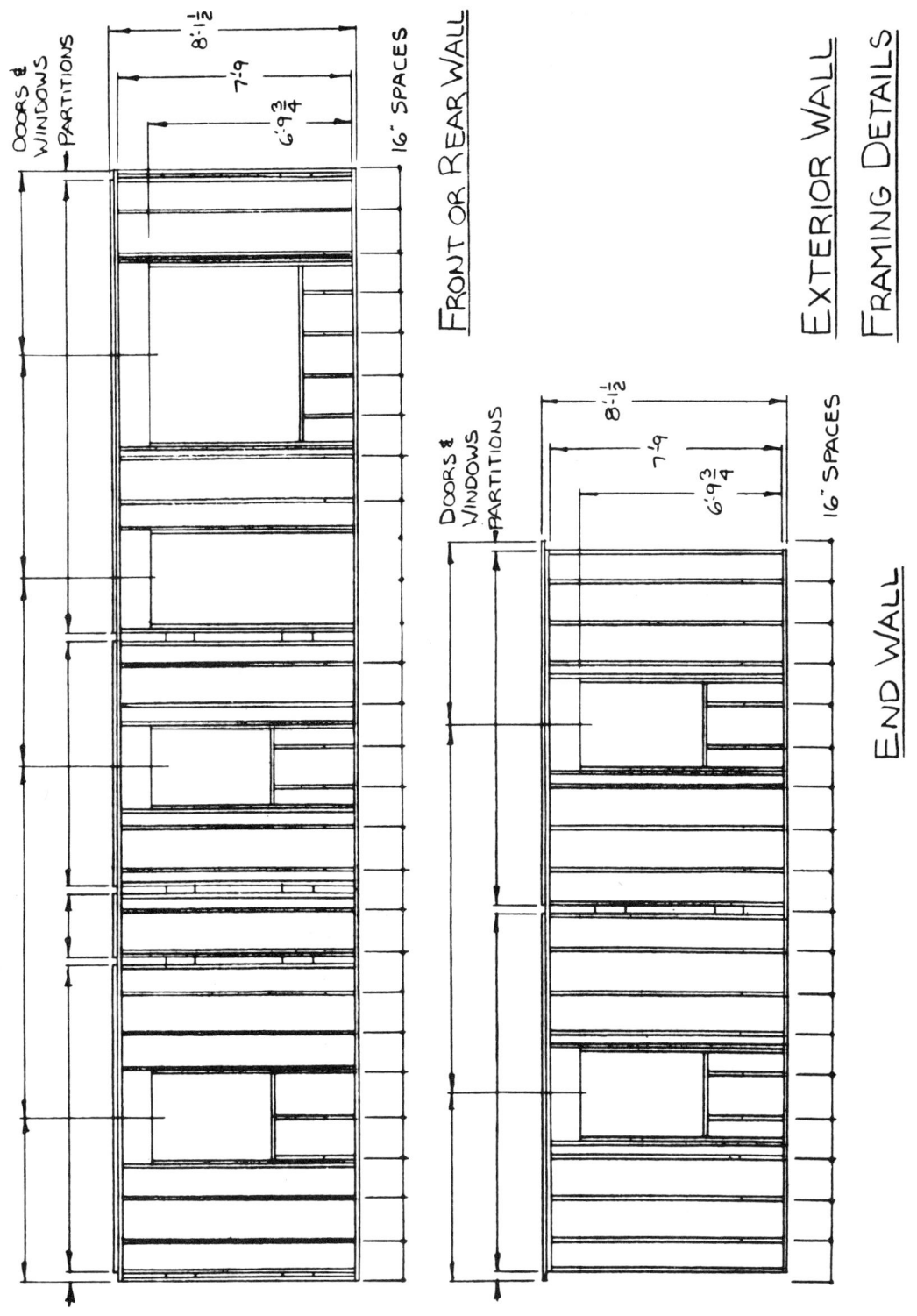

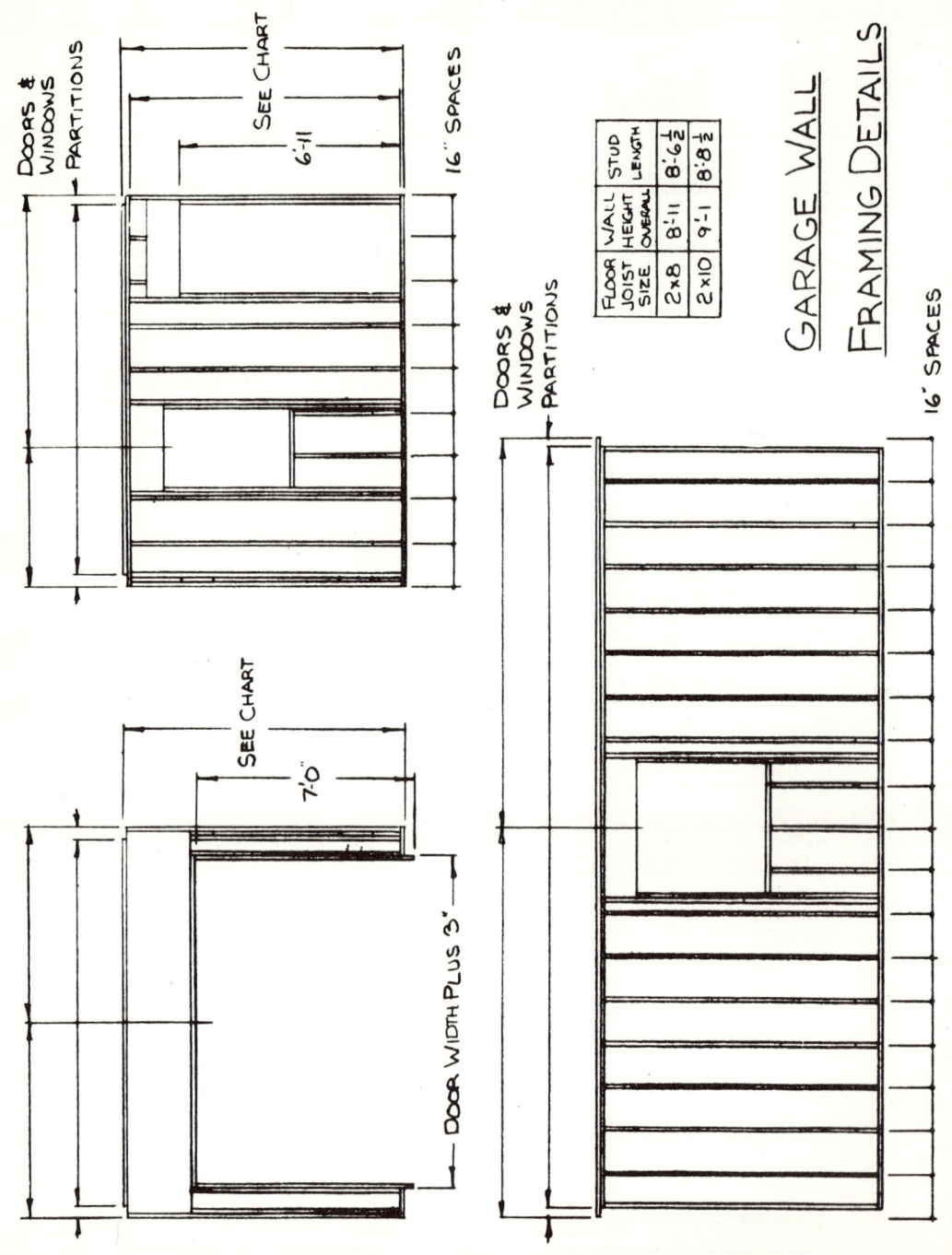

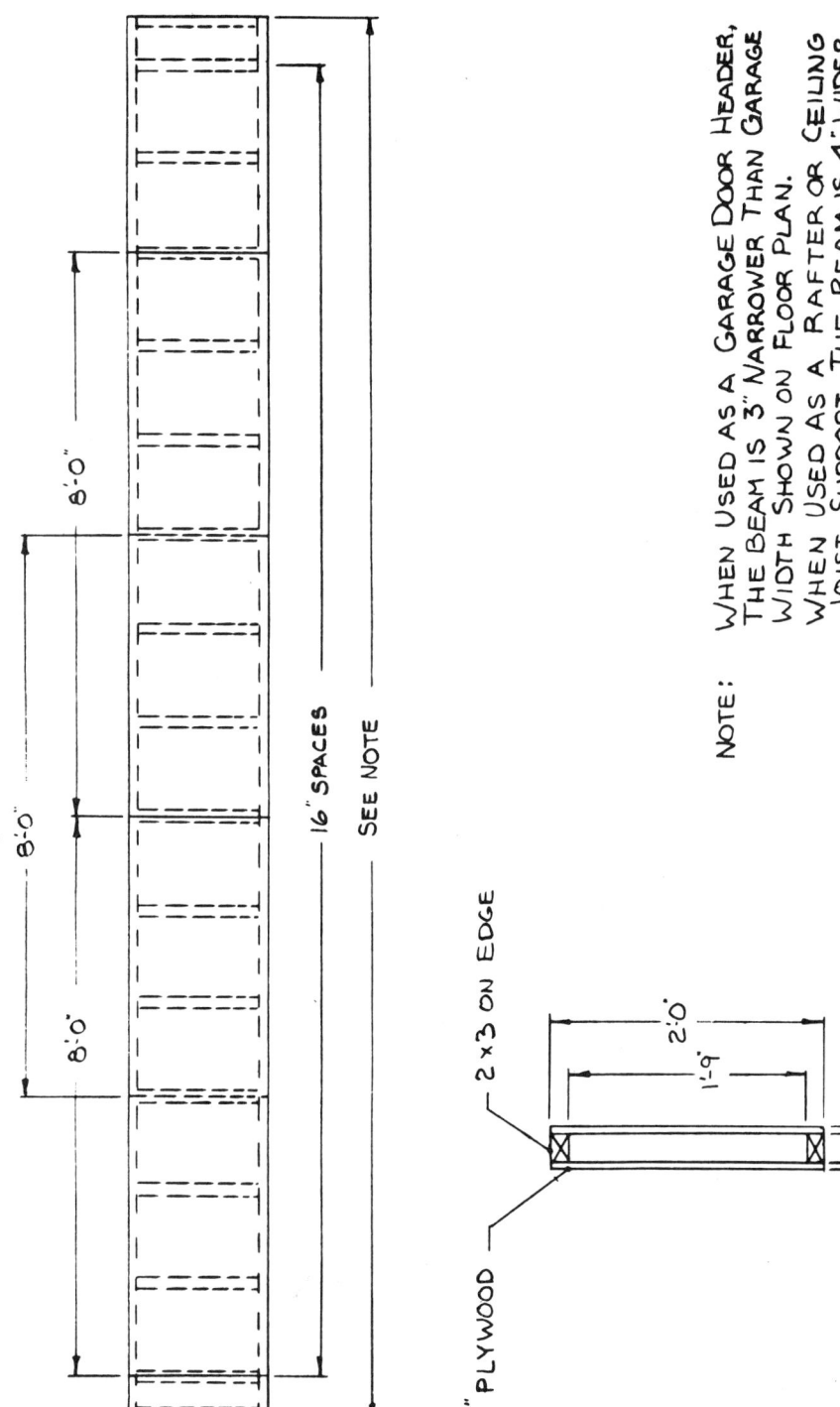

PARTITION
FRAMING DETAILS

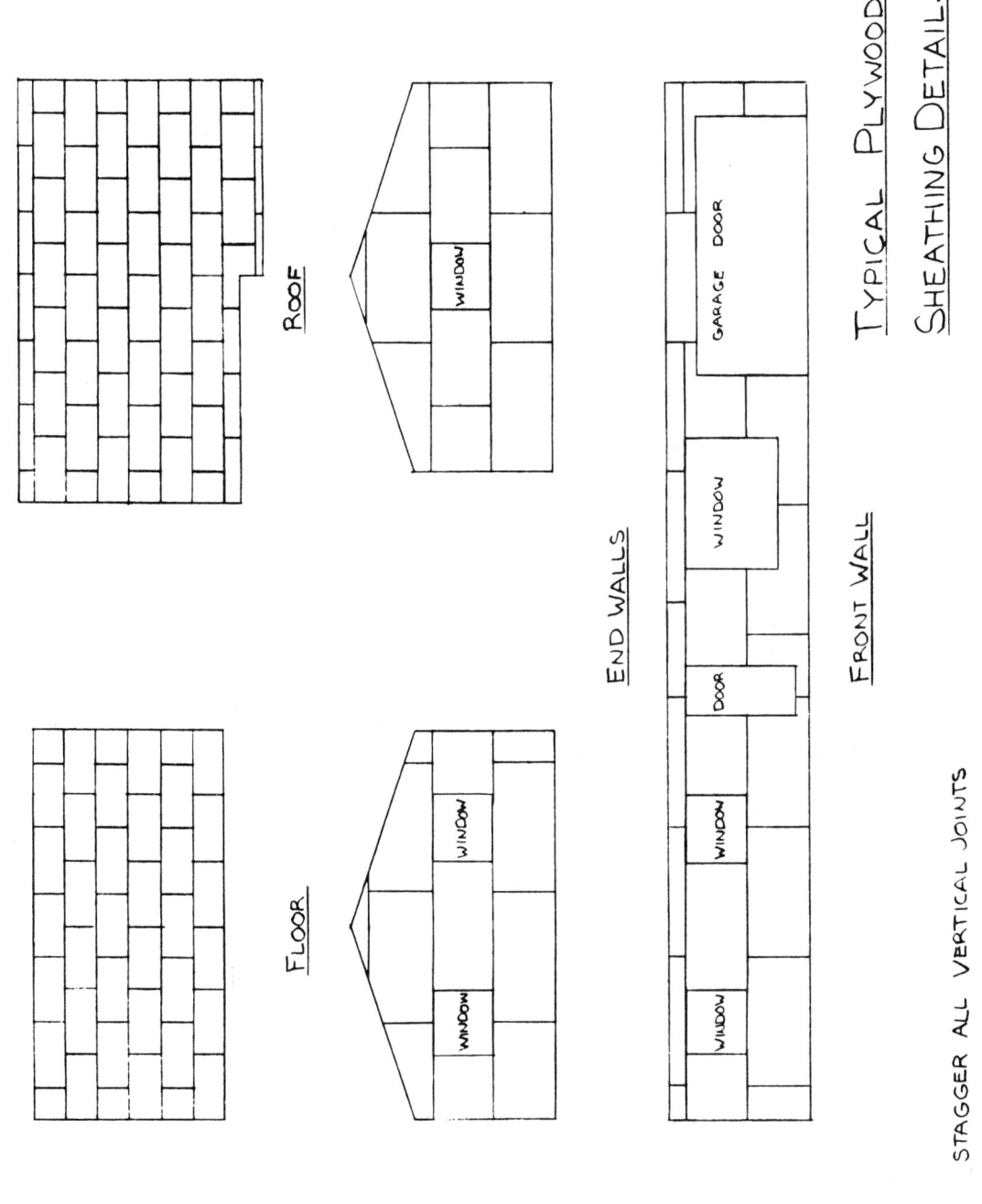

17. Roof Framing and Sheathing (4:1 Pitch Gable)

A gable roof is the easiest to build and the most commonly used, and is adaptable to almost any style. A 4:1 pitch allows a wide choice of roofings.

Using a steel tape, mark the location of the ceiling joists on the top of the supporting partition walls. Lay a 2 x 6 ceiling joist on top of the partition walls and toenail it to each outside top plate with three 10d nails. Join the joists over the center partition with a 2-foot-long 1 x 6 and nail with twelve 8d nails; toenail it to the partition top plate with four 10d nails.

If a portion of the center partition is left out and a flush ceiling will be used from front to back of the house, a beam in the attic space will support the ceiling. The beam can be made of two 2 x 12's nailed together and placed over the opening and supported by at least two joists, or an exterior wall on each end. The ceiling joists are hung from the beam by 16 gauge x 1 inch steel straps which are wrapped completely around both the joists and the beam and nailed to the beam.

Nail a 2 x 6 to the top of all partition walls which are parallel to the joists. This is to provide a nailing surface for the ceiling material. Use 10d nails staggered 16 inches on centers.

Cut the roof rafters accurately and lay on top of the ceiling joists. Cut the ridge board supports and nail to the ceiling joists at 8-foot intervals. Nail temporary bracing to hold them up. Cut the ridge board from very straight and true 2 x 8 lumber, in three or four long pieces.

Get help lifting the ridge board and rafters in place. Nail the first rafter of each pair with two 16d nails. Toenail the second rafter to

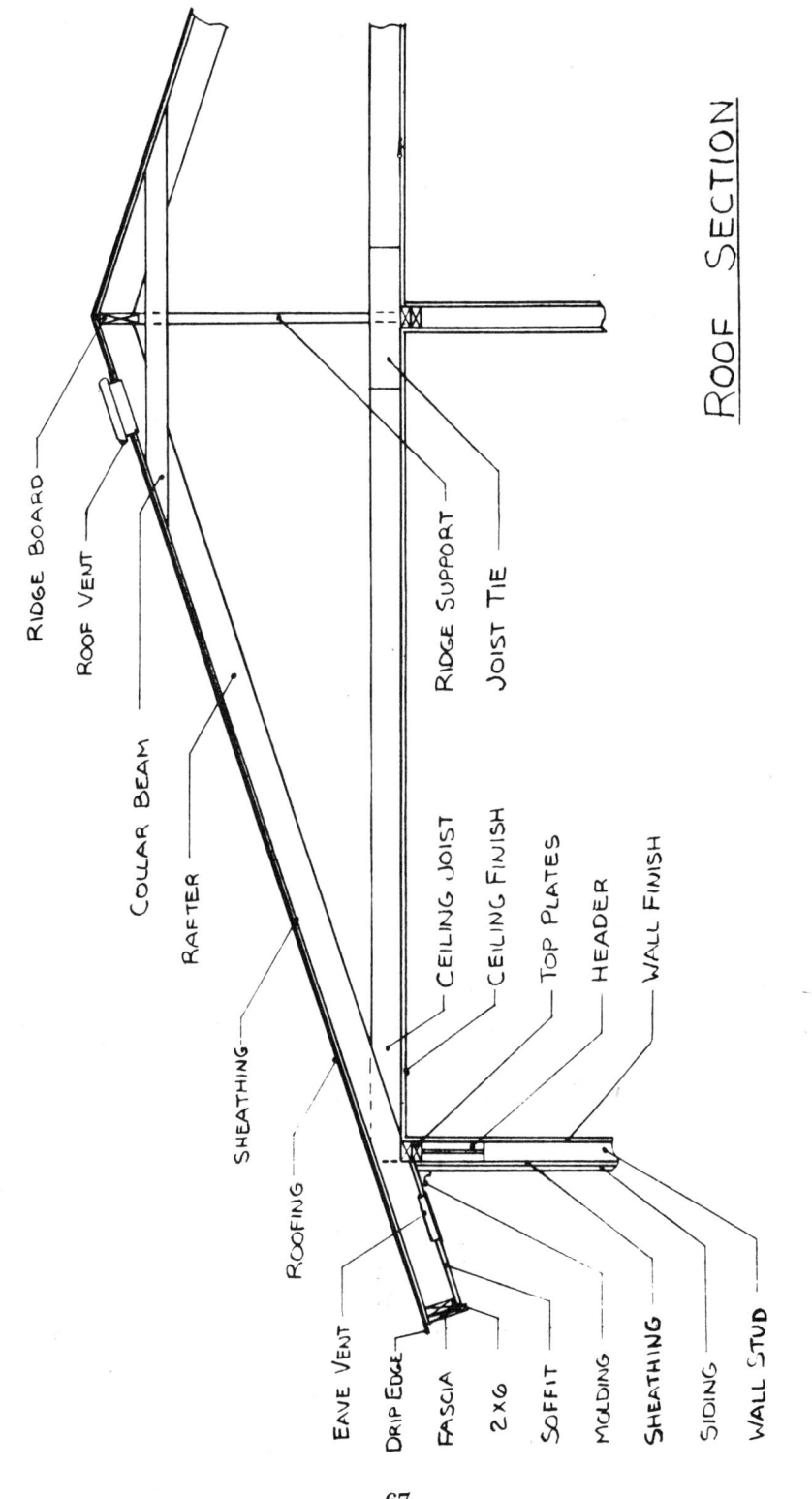

Depth of House	'A'	'B'	'C'	'D'	'E'
20'-0	10'-0	15'-8 1/2	12'-8 1/2	10'-5 3/4	3'-1 1/2
22'-0	11'-0	16'-9 1/8	13'-9 1/8	11'-6 3/8	3'-5 1/2
24'-0	12'-0	17'-9 3/4	14'-9 3/4	12'-7	3'-9 1/2

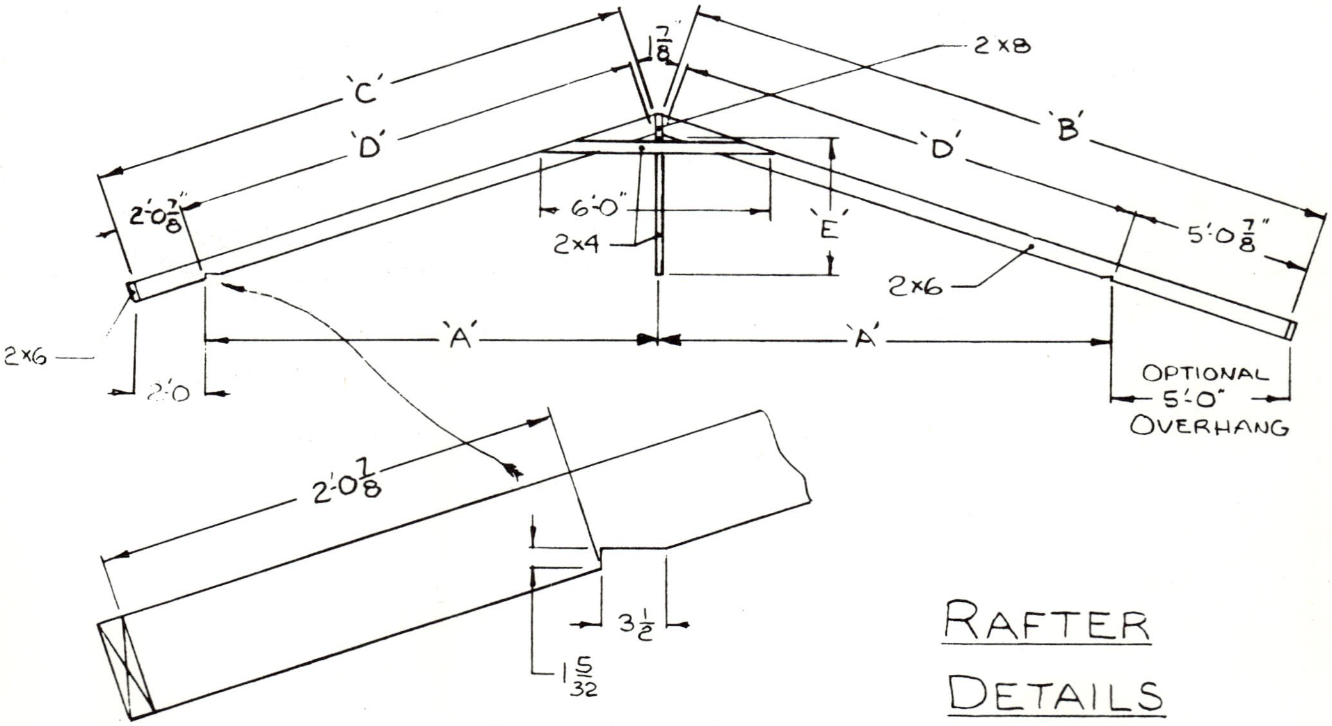

RAFTER DETAILS

20' TO 24'

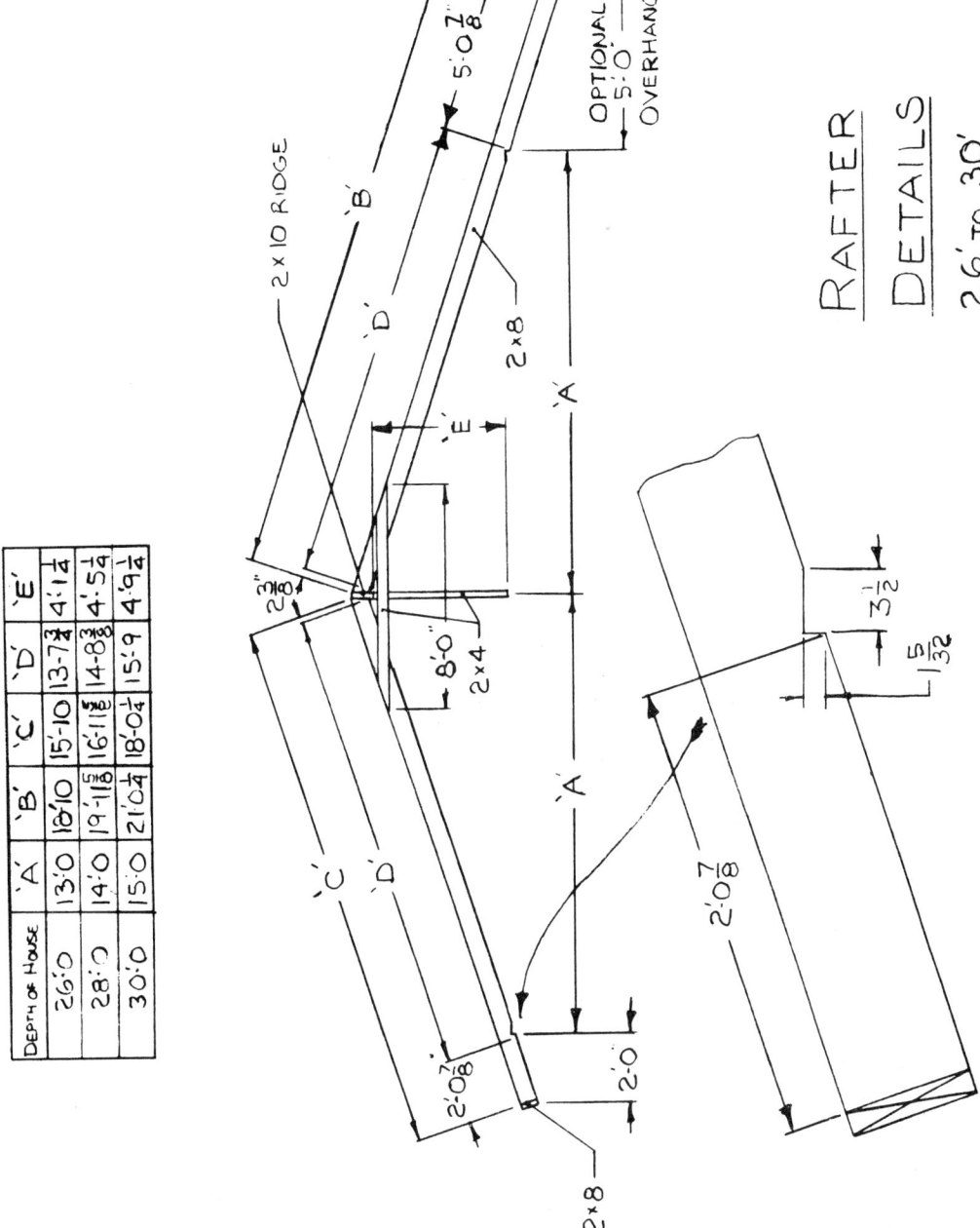

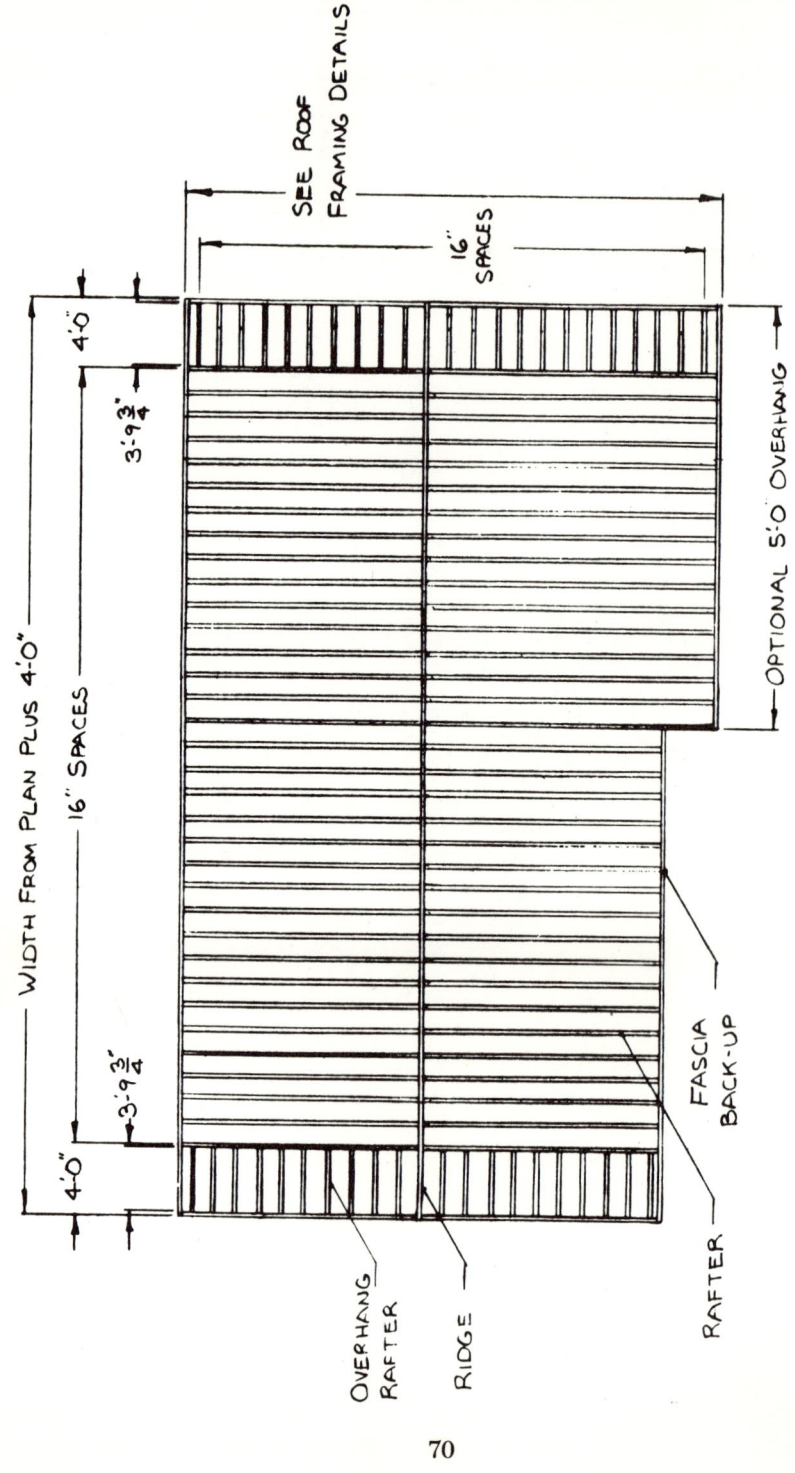

Roof and Ceiling Framing

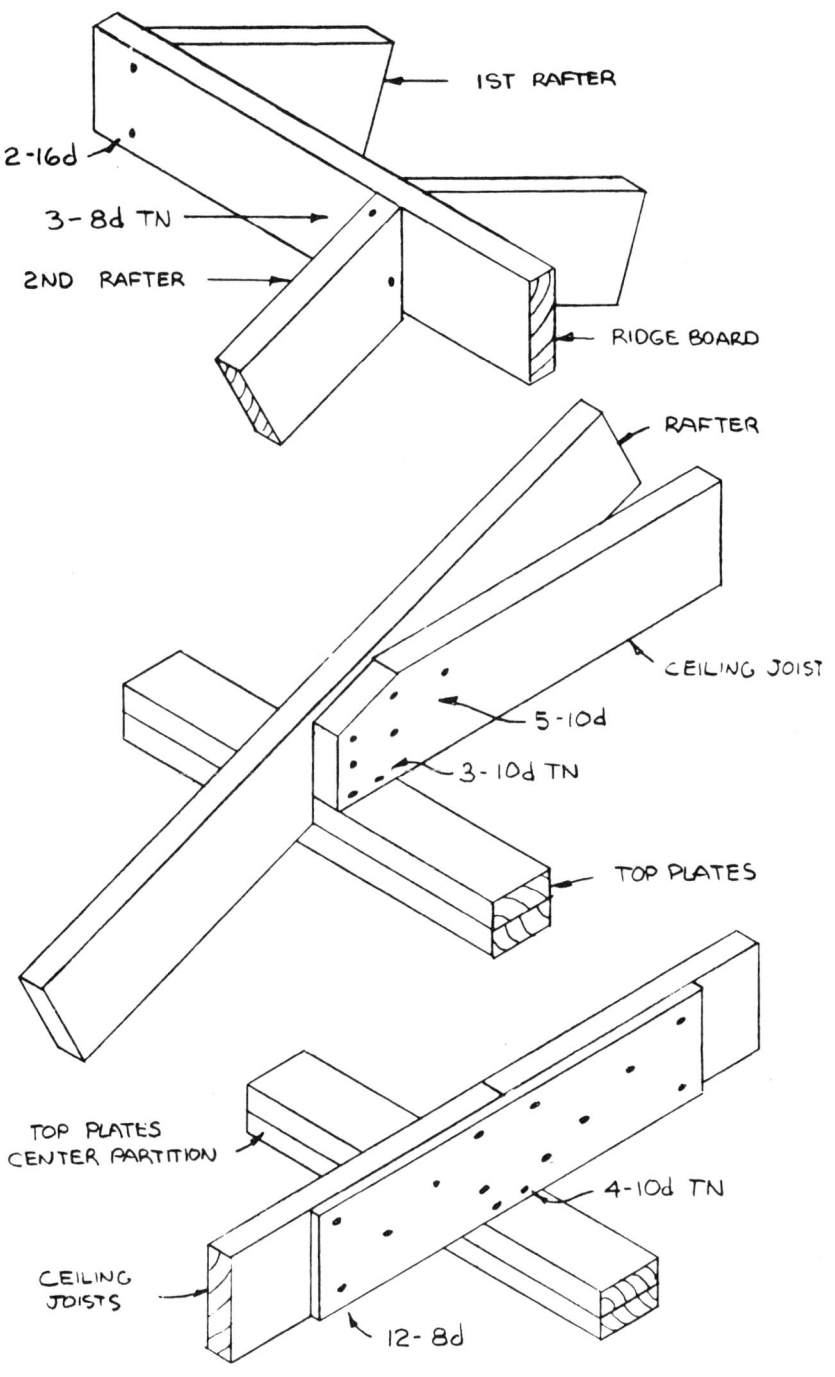

Nailing Surface for Ceiling Material

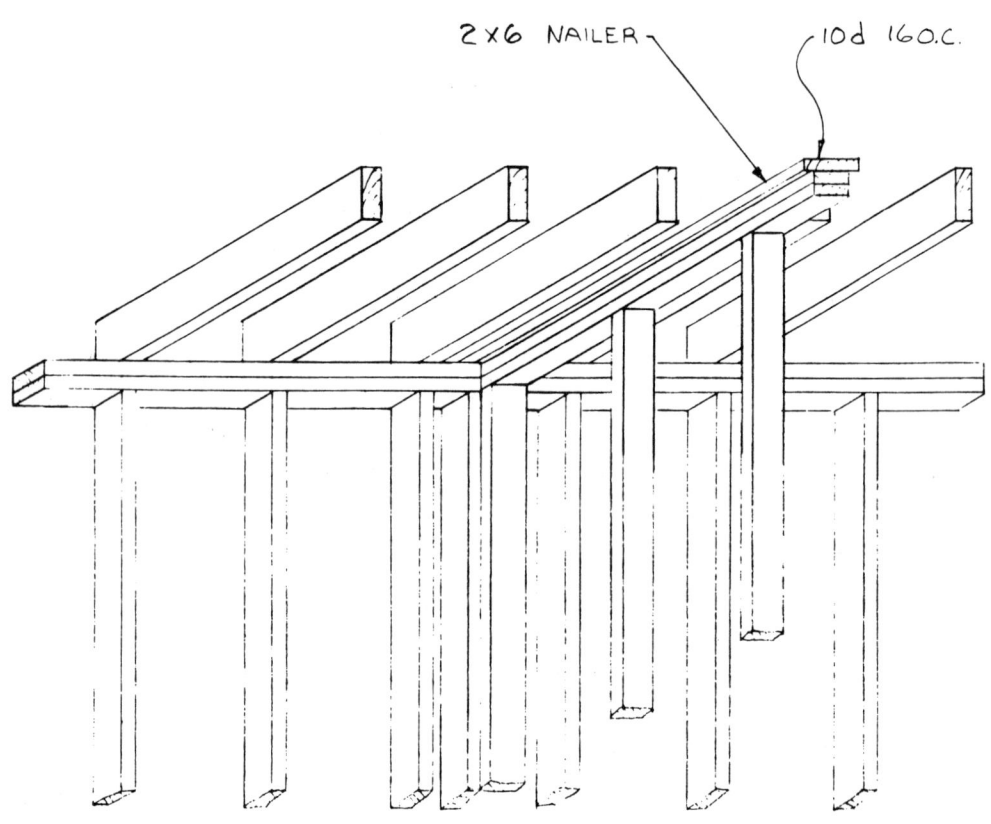

ROOF SHEATHING

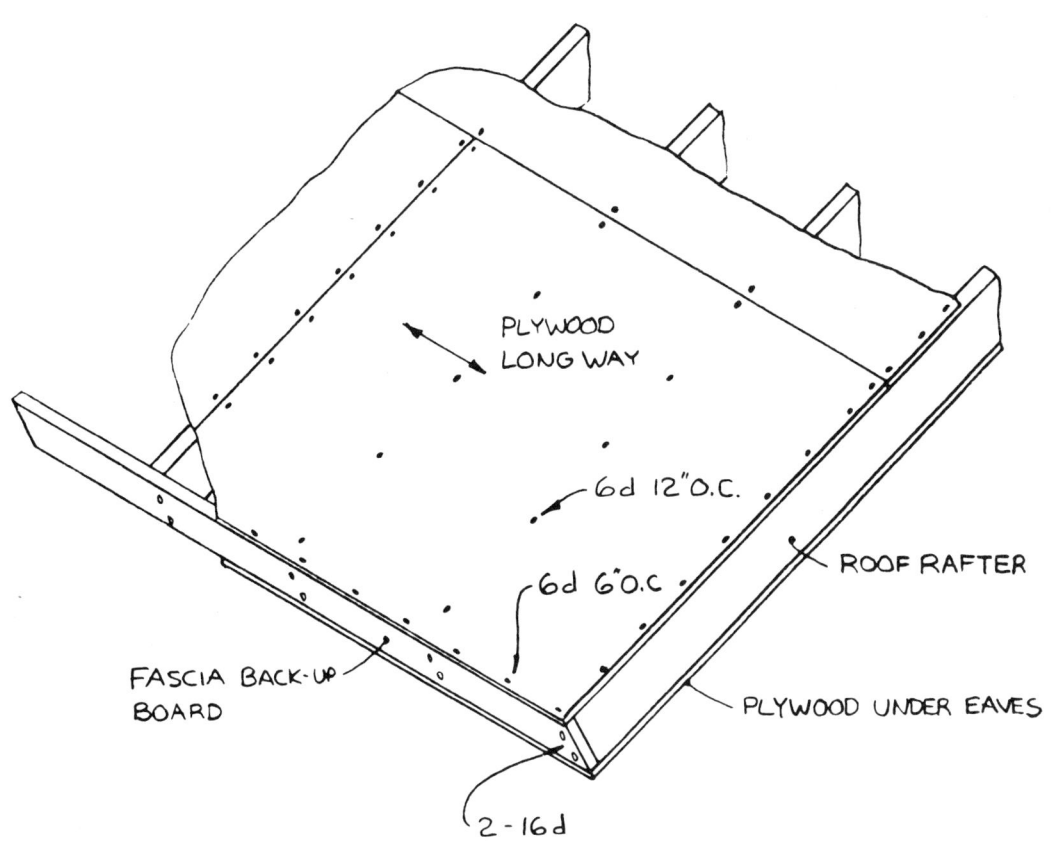

Depth of House	A	B	C	D	E
20	18'8	9'4	9'9 1/4	2'11 5/8	3'2 5/8
22	20'8	10'4	10'9 7/8	3'3 5/8	3'6 5/8
24	22'8	11'4	11'10 1/2	3'7 5/8	3'10 5/8
26	24'8	12'4	12'11 1/4	4'1 1/4	4'2 5/8
28	26'8	13'4	13'11 7/8	4'5 1/4	4'6 5/8
30	28'8	14'4	15'0 1/2	4'9 1/4	4'10 5/8

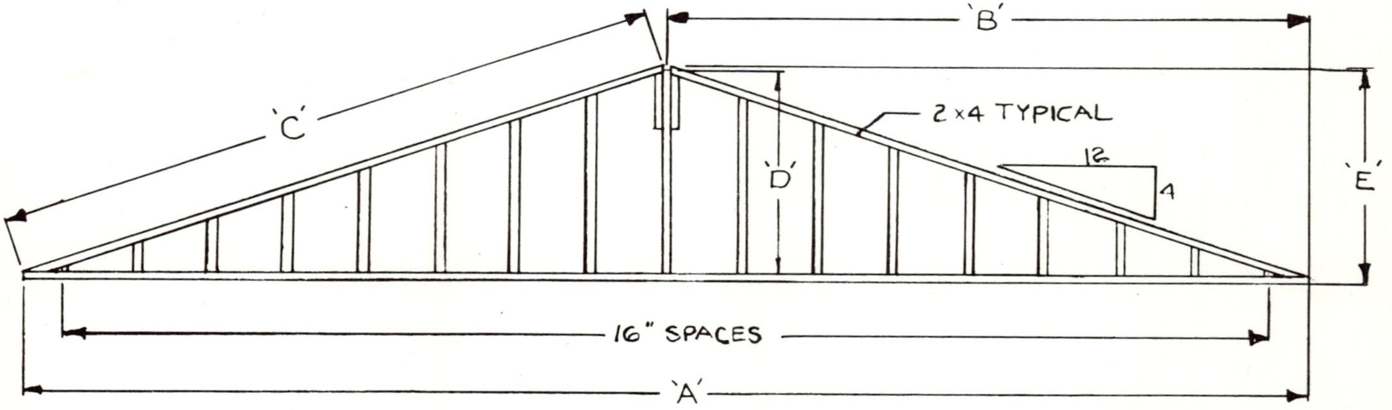

Gable End Detail

the ridge board with three 8d nails. Nail the rafters to the joists with at least five 16d nails. For houses wider than 26 feet, use 7 nails. Toenail rafters to the top plate with two 10d nails.

Cut the 2 x 4 collars and nail to every third pair of roof rafters with three 16d nails on each end. Nail the short overhang rafters on each end of the house with three 16d nails in each end.

The ends of the roof rafters are capped with a 2 x 6 fascia backup board. Use two 16d nails with each rafter.

The roof sheathing will be ⅜-inch exterior plywood. Use the pattern shown in the roof framing plan. Nail every 6 inches on edges and every 12 inches in the center with 6d nails.

Frame the gable ends as shown in the roof framing drawings. Nail the top plate to the roof rafters with two 16d nails per rafter. Then toenail the studs in place using two 10d nails on each end.

The garage roof is framed as part of the house roof, except that the garage has no center bearing partition. A built-up plywood beam is used to support the roof or ceiling joists. The beam is built of 2 x 3's with a 24-inch wide ½-inch plywood web on each side. Glue the beam together with waterproof resin glue and nail with 5d annular nails 6 inches apart on center. The beam is 4 inches longer than the garage width.

The garage ceiling can be eliminated if desired. Leave out the joists, and support the ridge board of the roof with the plywood beam. Build a support for the beam of two 2 x 4's and then back the support up with two more 2 x 4's nailed to the sides of the support and the beam. The outside end of the beam is supported by the gable. This will provide an open area under the roof which can be used to build storage space or to hang large articles.

If the garage is to have a ceiling, the beam can be supported by the top of the house ceiling joists and by the gable on the outside end. Hang the ceiling joists from the beam with 16 gauge by 1 inch steel straps wrapped around the joists and nailed to the beam. This will provide a flat unobstructed ceiling.

18. Roofing (4:1 Pitch)

Before starting roofing, install a metal drip edge (available where you buy roofing) over the edge of the roof sheathing. This will project about ½ inch over the edge of the fascia and will support the roofing.

Eave flashing is required where the outside temperatures fall below 0°F. This requirement is met by using a double layer of underlayment from the eaves to at least two feet inside the exterior walls. Roofing underlayment is 15- or 30-pound saturated felt. For a 4:1 pitch roof, one layer of underlayment is required over the entire roof. Use as few nails as necessary to hold it in place until the shingles are installed.

The most common roofing is asphalt shingles. The best kind has a dab of cement on the underside of each tab which seals the shingles together with the heat of the sun. This makes it almost impossible for them to blow off. Most asphalt shingles are double coverage. Nail according to the manufacturer's instructions. Double up the first course. Use 11 or 12 gauge corrosion-resistant roofing nails. They must be long enough to go all the way through the roof sheathing.

Type of Shingles	Shingle Length	Maximum Exposure
Wood shingles	16 inches	4½ inches
	18 inches	5 inches
	24 inches	6¾ inches
Wood shakes (hand split)	18 inches	8 inches
	24 inches	10 inches
	32 inches	13 inches

Wood shingles are also very popular, because of their rustic appearance. They must be laid with the correct exposure, as shown

ROOFING

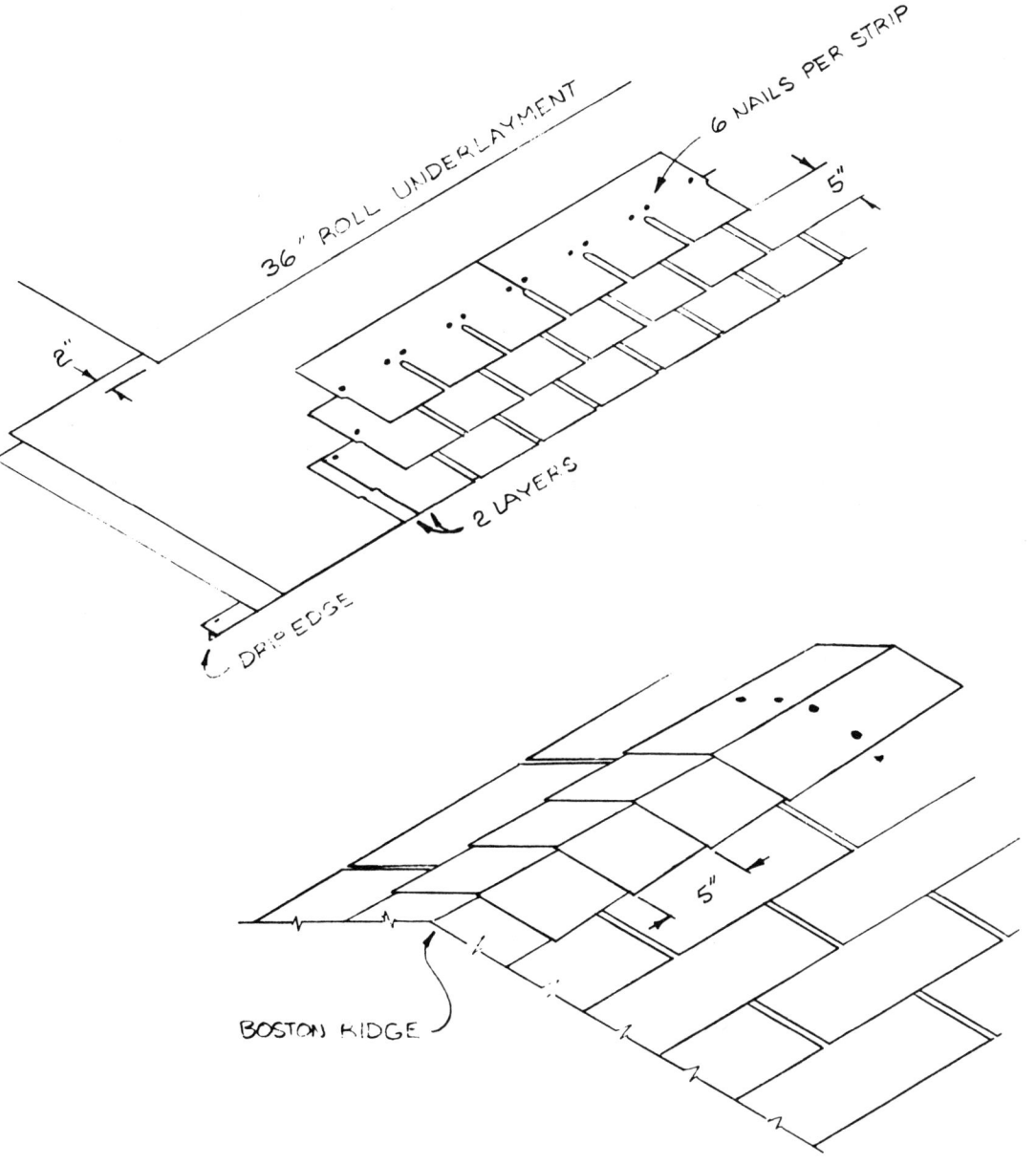

ASPHALT SHINGLES

above. These shingles are made of Red Cedar, Cypress or Redwood. They are laid approximately ¼ inch apart to allow for expansion. Be sure all joints overlap for three courses. Use two corrosion-resistant wood-shingle nails for each shingle. They should be the threaded type, and go all the way through the roof sheathing. The nails should be covered 1 to 1½ inches by the next course.

When laying shingles, it is advisable to use a chalk line to mark parallel horizontal lines about two feet apart on the underlayment, as a reference to assure even rows. Adjust the spacing so that the top course will come out even. Finish the ridge by laying the shingles sideways to form a Boston ridge. For wood shingles, overlap the ridge with every other shingle to the opposite side. Sometimes every sixth course of wood shingles is doubled to make a shadow line.

Be sure to provide roof ventilation near the ridge. (See *Flashings, Vents, and Gutters*.) Also, if possible, provide the flashings for the soil stack at this time.

19. Windows

There are many kinds of windows available. Double hung windows are the most common and therefore the most perfected in design and lowest in price. Look for windows with aluminum tracks which are spring-loaded to allow the sash to be easily removed for cleaning. These also have spring lifts to hold the sash up. Aluminum triple-track storm and screen windows are available to fit these windows and provide ideal weather protection.

Many other types of windows are available, such as sliding or gliding windows, casements which swing out like a door, hopper windows which hinge in from the bottom and vent windows which swing out from the top.

Special window units are also available for special purposes. Bay or bow windows project out from the house. Some types open and others are fixed. Picture windows are large windows and do not open, although sometimes they are flanked by smaller windows which do open. Window walls are groups of large windows which cover the greater portion of a wall.

Storm windows are available for all types of windows, but casement and vent windows usually have the storm sash fastened to the window sash, so good weather stripping is required. All windows are available with double glazing such as Thermopane. This eliminates the need for changing storm windows and cuts in half the glass surface requiring cleaning. Good weather stripping is required here too. Triple-track aluminum storm windows also eliminate the need for removing and storing the storms and screens each year: the windows slide up and down easily to change. (See *Flashings, Vents, and Gutters* for details of top flashing for windows.)

Any type of windows can be used, but try to use only one kind and as few different sizes as possible to achieve a planned look.

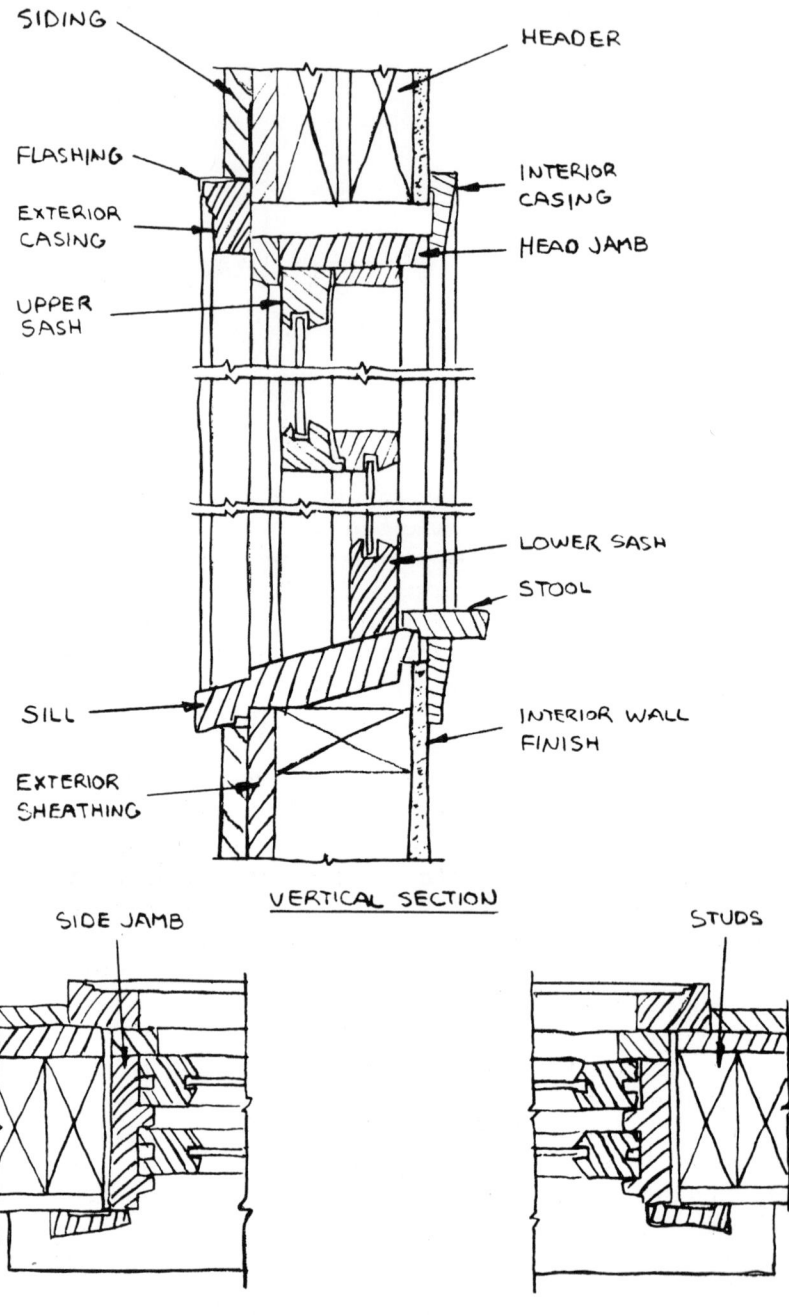

WINDOWS

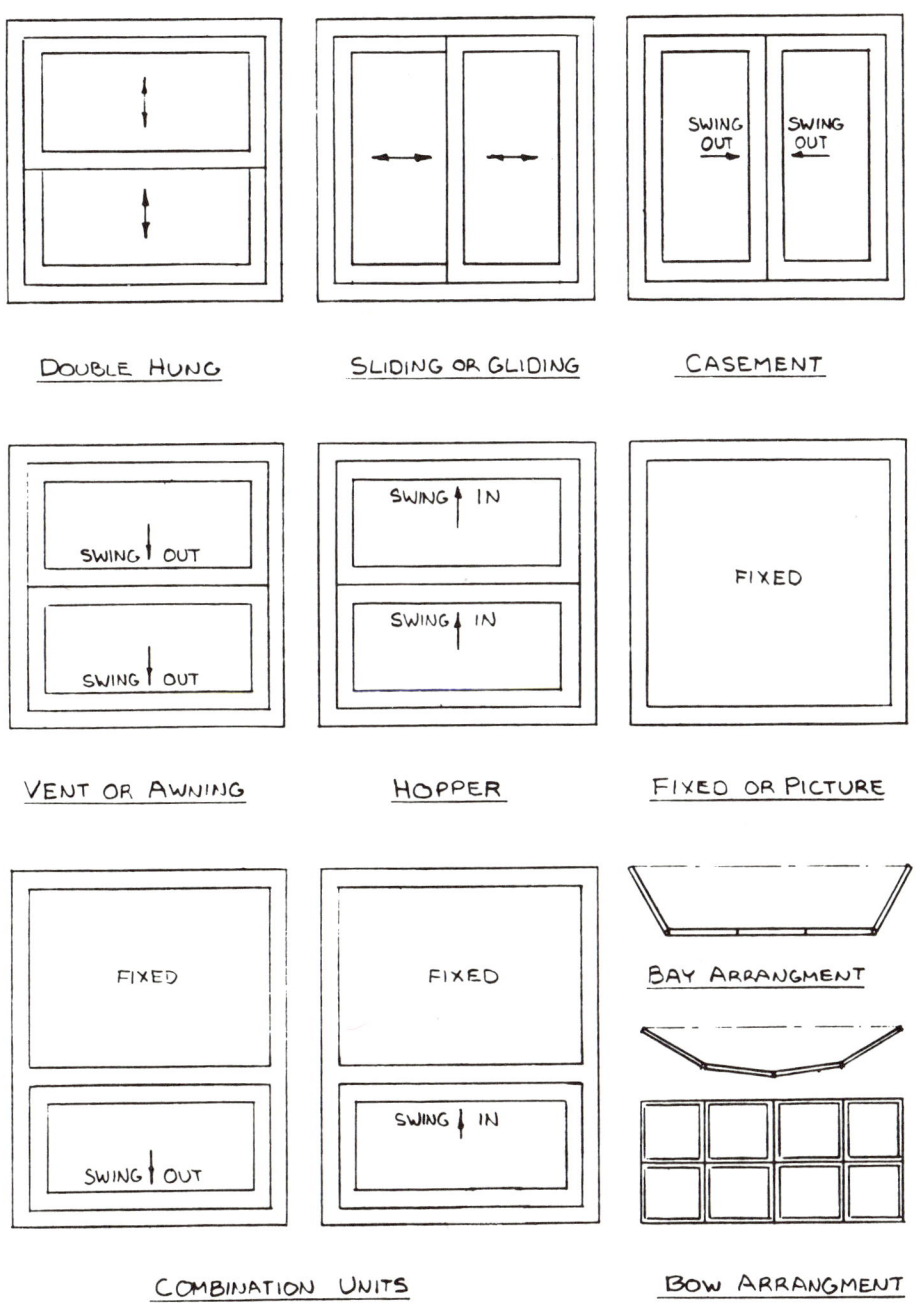

Get accurate rough opening measurements from your window dealer because many window manufacturers use nonstandard sizes. The tops of all the window rough openings will be the 2 x 12 headers.

Your windows should come with the exterior trim and window stools installed. Where two windows are to be installed side by side, they can be purchased as a pair with the exterior trim in place. When buying windows, the correct jamb width must be purchased. Add together the 3½-inch stud width, the exterior sheathing thickness and the interior wall finish thickness to determine the correct jamb width to buy. It is also practical to buy the windows with the primer coat on them. They are available with vinyl covering, but these are quite expensive. If the windows do not come primed, they should be primed before they are installed.

Place the window in the opening and shim up until the jamb extensions are against the top header. You can use wood shingles for shims around the window. Be sure the window is level and solidly shimmed, then nail through the exterior trim into the house framing all around. Use finishing nails long enough to penetrate the framing by 1 inch.

20. Doors

Doors are available in either flush style or panel style. Exterior flush doors are available with a variety of window designs. Interior flush doors are usually made of mahogany or birch; panel doors are usually pine, and are available in a wide variety of designs, with windows for exterior use. They are also available in many designs for interior use. Closet doors can be the louvered type, which are very attractive but hard to finish and they gather dust easily. Bifold-type doors are recommended for closets because they do not require any jambs on the sides, and open to reveal the entire closet. Panel doors are more expensive than flush doors but are becoming very popular and fit in well with either rustic or formal designs. Exterior doors are 1¾ inches thick and closet doors can be 1⅜ or 1⅛ inches thick.

Sliding patio doors are available in either wood or aluminum. The aluminum doors are very strong but vary widely in quality. Be sure to get doors with good weather stripping, especially in cold climates. Also, get double glazing, and tempered glass. Patio doors are very nice for rooms which open onto a private yard, and can be used in any room.

Exterior door frames should be purchased with weather stripping and exterior trim installed, if possible. When installing exterior door frames, the floor sheathing and sill are cut out to allow the door sill to be flush with the level of the finished floor. Level the door frame by using wood shingles as shims and nail through the trim into the house framing. Use finishing nails long enough to penetrate the framing by at least 1 inch. Then nail through the jambs into the studs for extra support. Interior doors are hung the same way except that they are nailed through the jambs only. Prehung doors are also avilable to cut the installation time.

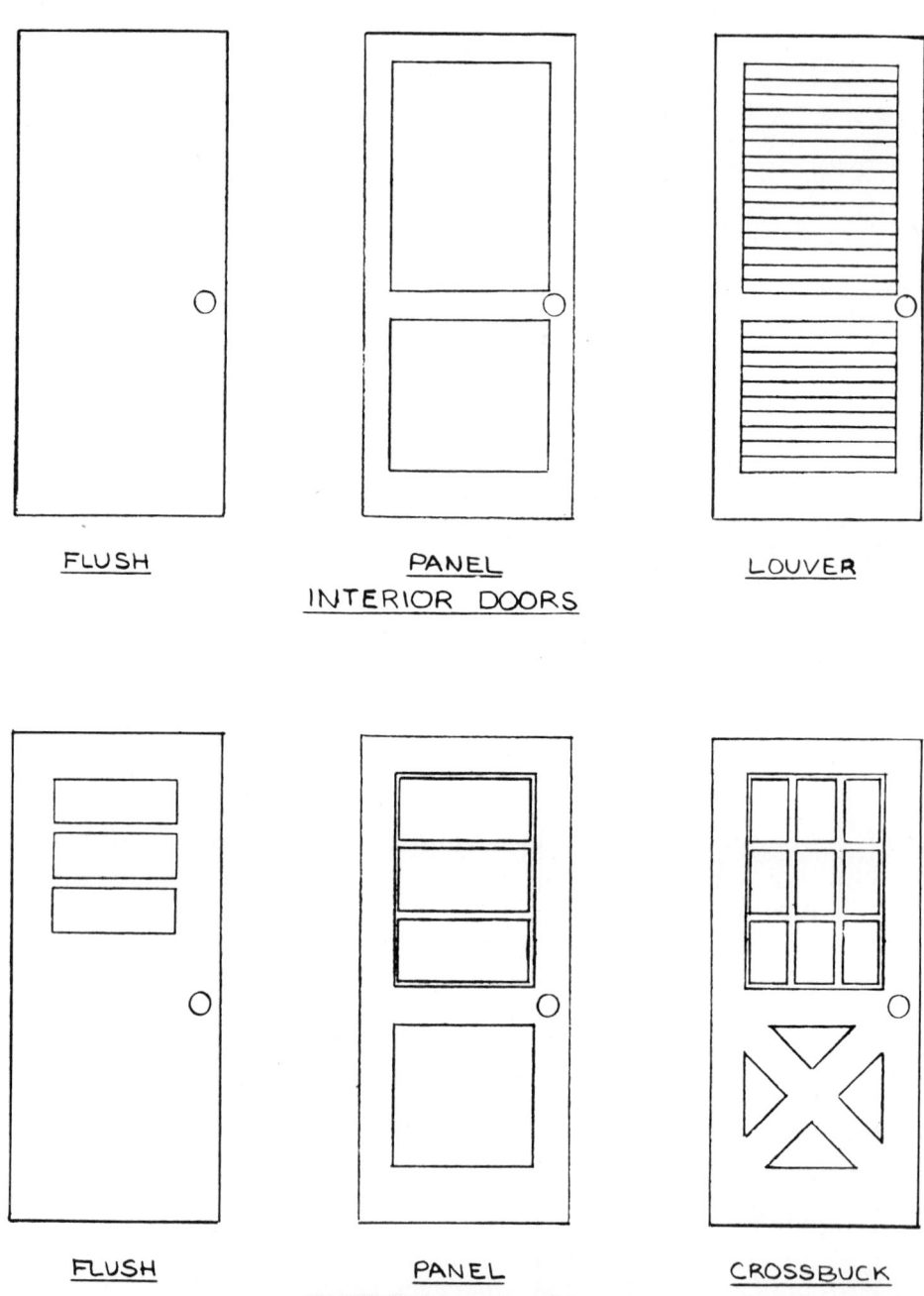

EXTERIOR DOOR DETAILS

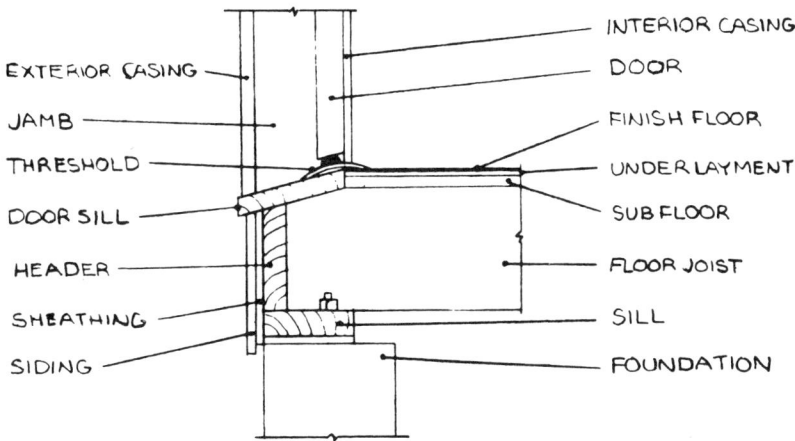

Storm and screen doors may be the wood type which are installed in the door frame, or the prehung aluminum type which are simply screwed to the frame. Aluminum screen doors are available with a baked enamel finish if the aluminum finish is not desired.

Exterior doors are hung with three hinges; interior doors, with two hinges. Drive a wood block between the jamb and the framing opposite each hinge or strike plate. Measure down 5 inches to the top of the top hinge, and measure up 7 inches to the bottom of the bottom hinge. Center the third hinge for exterior doors. Chisel an area the depth of the hinge and 1⅛ inches wide for interior doors or 1⅜ inches wide for exterior doors into both the jamb and door to place the hinge. The hinges should be 3½ x 3½ inches for interior doors and 4 x 4 inches for exterior doors. Sliding or bifold doors are hung with special hardware. Locks and door knobs must be installed on all doors except closets. Exterior doors require key locks, bathroom doors require privacy locks, and other doors require just knobs and latches. Use the bored-in type, which can be installed by drilling two holes, one for the knob and one for the latch. A special boring tool, rented or borrowed from your dealer, will do the job easily. Aluminum thresholds with waterproof vinyl inserts are used under exterior doors.

21. Flashings, Vents, and Gutters

Flashings must be provided in most areas to keep out moisture. This is extremely important. Use an approved corrosion-resistant flashing material such as copper, aluminum, or galvanized steel. Your dealer can recommend what is used in your area.

After the windows and doors are installed and before the siding or brick veneer is applied, a metal flashing must be installed over all windows and doors. This flashing should cover the top of the door or window trim and extend ½ inch down and 2 inches up.

If a combination of building materials is used, such as brick veneer up to the window sills and siding above, the intersection must be weather-stripped with a flashing extending two inches up and across the top of any trim. If the trim is stone or brick, extend the flashing out flush with the siding and caulk the joint between the flashing and the brick.

Roof flashing is used over the edge of the roof. (See *Roofing*.)

Attics and crawl spaces must be provided with adequate ventilation. For crawl spaces, install ventilators having a total free-ventilating area of at least one square foot. Cut holes through the siding and floor header to accept these ventilators. If the wall is brick, cut the hole through the header and use a special ventilator which is put in place when the brick is laid. If you are going to have a brickmason do this, he may be able to provide the ventilators.

In attics the vents are under the eaves and along the ridge of the roof. The ridge vents are installed by cutting holes in the roof sheathing about one foot down from the ridge on the back side of the house. It is best to install these at the same time the roofing is done. Use enough ventilators to provide one square foot of free ventilation for each 600 square feet of floor area. The ventilators are made of aluminum or plastic. Plastic vents which match the

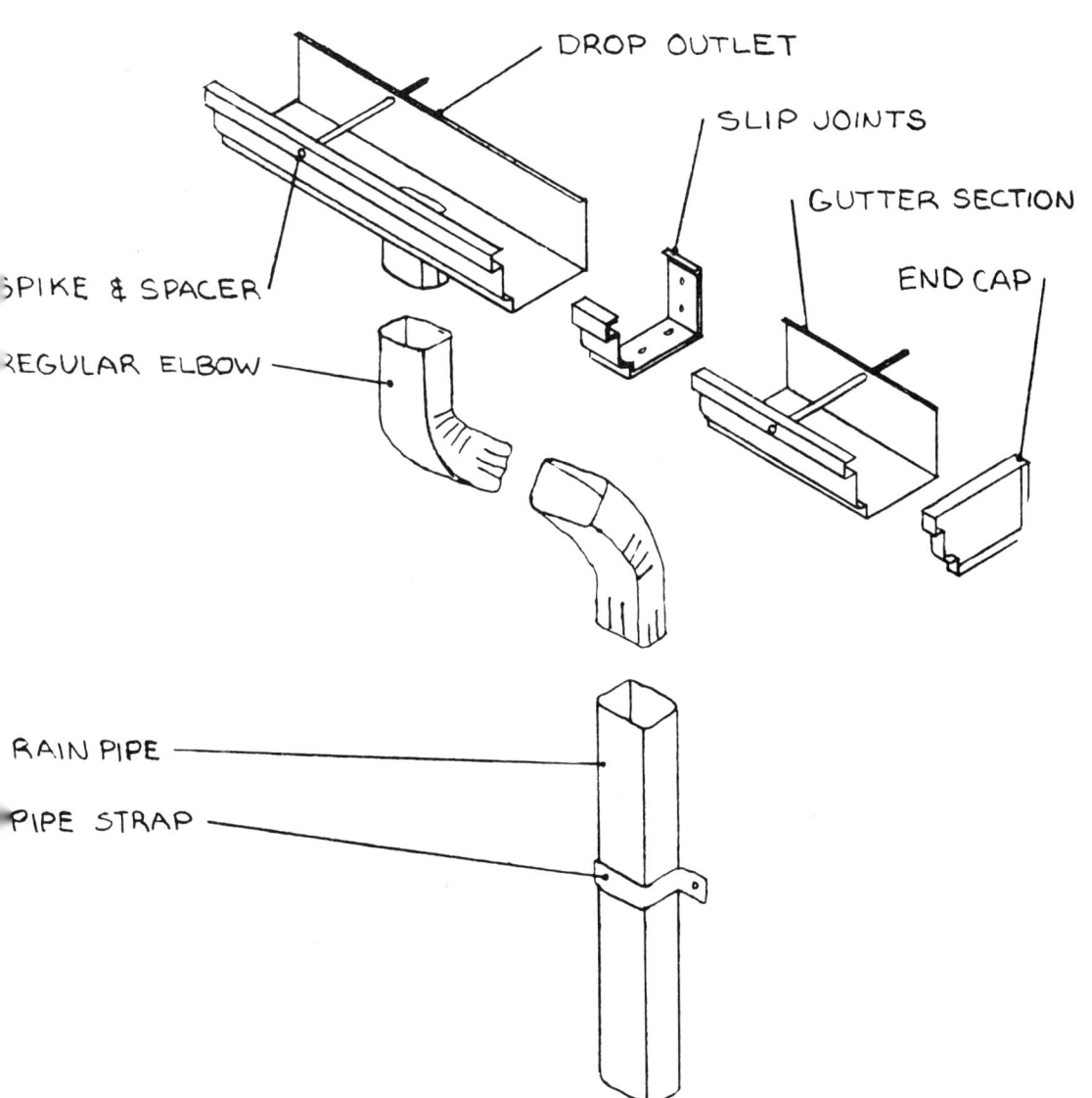

roof color are available. The eave vents are installed over openings cut in the underside of the eaves on both sides of the house. If aluminum will be used to cover the underside of the eaves, it is available with perforations for built-in ventilation. Space all ventilators evenly for a neat appearance.

Vents are also necessary for the bathroom vent fans, kitchen exhaust fan, and automatic clothes dryer if you will have one. Usually 4-inch-diameter aluminum pipe is used for these. The ventilators come in three types. One vents through the roof, and another, usually used with clothes dryers, vents through the wall. If the dryer will be in the basement, this one will go through the floor header. Be sure to leave a hole in the wall for this if brick is used. The other type of ventilator goes through the underside of the eaves and should be used for baths and kitchen for the neatest appearance. Flexible pipe is available for easy installation.

Gutters and downspouts are usually used if the house has a basement, and in some areas must be used for crawl spaces or slabs as well. If houses in your area, of the type you plan to build, have gutters, you should probably also have them. Gutters are available in two styles—rounded or box type. Use the box type. These are available in aluminum and galvanized steel, unfinished or with white enamel finish. The aluminum gutters are far better, but cost slightly more. If you plan to paint them, the galvanized type is all right, but let them age for a year before painting. The gutter sections are connected by slip joints which must be filled with a special gutter caulking before assembling. Use a caulking gun and completely fill the joints. The sections are usually attached to the fascia board with spikes and spacers. Drill holes in the sections every 32 inches for the spikes, install the spacers, and nail to the fascia. Install one section at a time. Slope the gutters about ½ inch from the center of the house to the ends. End caps and drop outlet sections are used at the ends. The downspouts are installed at each end of the house, even with the ends of the house. Use elbows to bring the rain pipe flush with the side of the house and attach it to the house with two straps.

22. Exterior Siding and Trim

Choose the siding material you prefer. The trend in modern home design is toward low-maintenance sidings. If brick is not used, aluminum, steel, or vinyl siding will provide a very low maintenance exterior. The latest trend is toward the more rustic appearance of stained siding. This is usually cedar or redwood that is rough-sawn and pressure-treated or soaked to provide a lasting stained surface. When using either aluminum or stained wood, the trim can be done in the same material, if desired. Sometimes plywood is used under the eaves to reduce costs; it has to be painted, but in this protected area the paint lasts a long time. If you do not mind painting, there are many kinds of wood, pressed board, hardboard and plywood sidings available. With the new water-base paints, the painting is not too much of a chore, especially on low ranch-type homes.

Sidings are available in many types. The most common is the regular horizontal bevel type, which comes in many widths. This can be installed over ⅜-inch plywood sheathing by nailing into the studs.

Plywood siding comes in sheets that have grooves cut in them to give the appearance of vertical siding or rough-sawn textures. Sheets can be nailed directly to the studs in place of the sheathing if the climate and building codes allow. If not, it can be nailed over the ⅜-inch plywood sheathing.

Vertical siding, which comes in many types and widths, is becoming very popular. Board and batten is also very popular. Plain boards are used and the joints are covered with narrow wood strips. Vertical siding may be nailed to ⅜-inch or ½-inch plywood sheathing. An ideal siding for the owner-builder is pressure-stained cedar of the board and batten type, with matching trim.

If aluminum, steel or vinyl siding is used, be sure to get a good

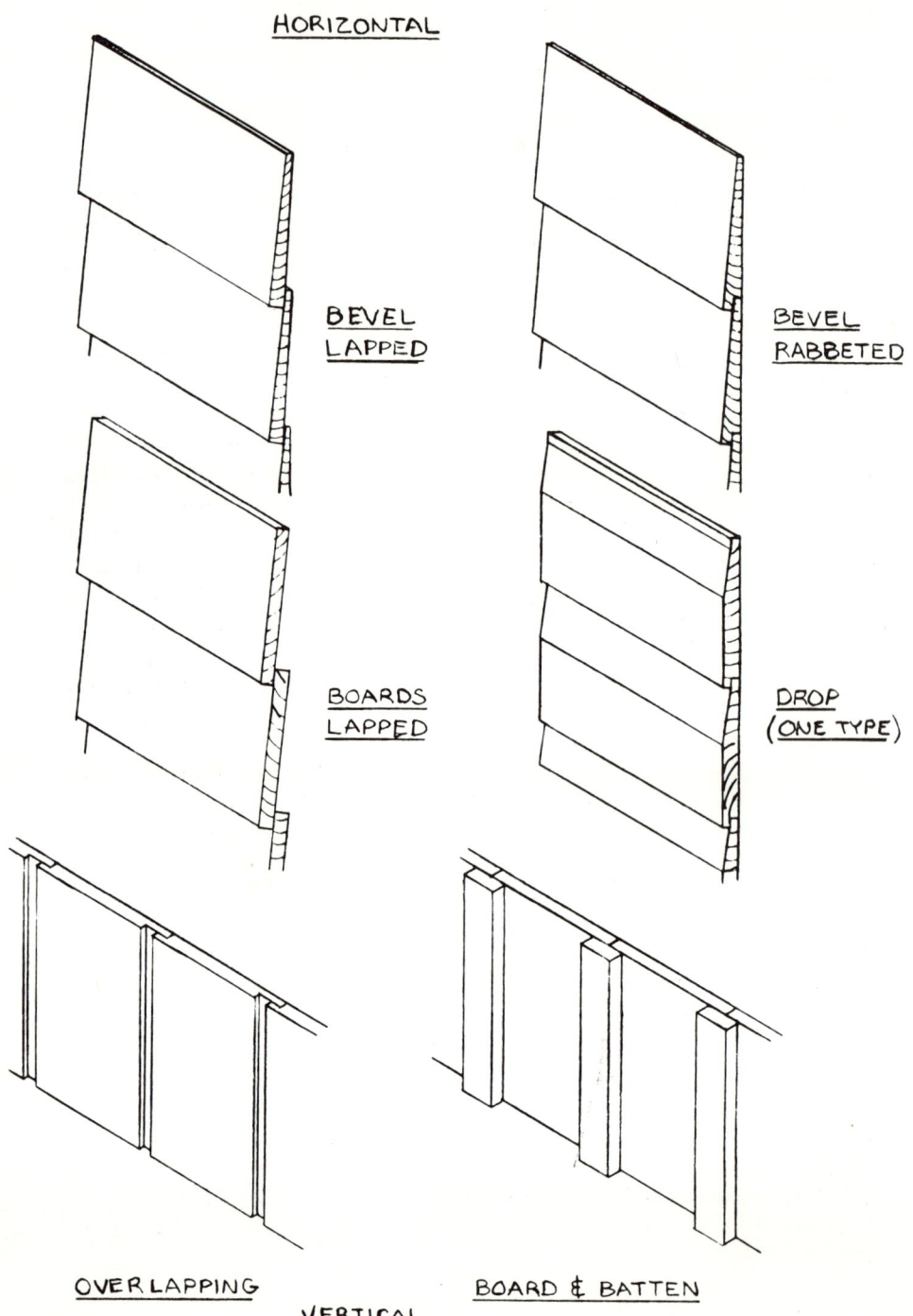

grade with a strong backing which will not dent. It is very easy to install by following the manufacturer's directions. Special shapes are available for use as trim. These cover the undersides of the eaves, the fascia and the door and window trim.

Wood shingles or shakes can also be used as siding. They are overlapped at least half their length, and may be nailed to ⅜-inch or ½-inch plywood sheathing. Use a double (or starter) course at the bottom of the wall.

All siding should be nailed with corrosion-resistant nails long enough to penetrate at least 1 inch into the studs. If the siding is nailed to the sheathing, use threaded or annular nails long enough to go all the way through the sheathing.

Many other kinds of siding can be used, such as diagonal, asbestos-cement and stucco. Follow the manufacturer's recommendations for installing them.

23. Plumbing and Drainage

Plumbing requires no special skills and only a very rough knowledge of the principles involved. Any do-it-yourself builder should seriously consider doing the plumbing himself. The two main objectives of a drainage system are to drain all the water and waste downhill, and let air in from a roof vent to keep the water from being held back by suction.

If your property has sewers, hire a plumbing contractor to bring the sewer and water to your house. Most areas require a licensed plumber to attach your plumbing to city water and sewer systems, which is a very important job. If it is not done correctly, the cost of redoing it can be very high. The contractor will dig a trench up to your foundation and run the water and sewer lines under the footings. Try to make an agreement on where the sewer line will enter the house. If possible have the sewer run directly toward the soil stack. See your floor plan to locate the soil stack, which is in the wall directly behind one toilet. Sewer and water may be brought in at any time from excavation to the time plumbing is done. The sewer pipe should be steel or iron, rather than the clay pipe sometimes used. The water pipe is almost always ¾-inch or 1-inch copper tubing. Run the copper water pipe to the point where you will have the shutoff valve. Have the plumbing contractor leave the required length when he connects to the city water.

Shop around for a good plumbing supplier. Some work only with licensed plumbers and are very uncooperative with amateurs, but there is usually one supplier who will work with a do-it-yourselfer and even give free advice. Sears, Roebuck and Montgomery Ward are excellent sources, although their prices might not be the lowest. They also have a plumbing design service.

Under the floor of the basement, slab, or crawl space is the building drain. This is the pipe from the soil stack to the sewer. There are four types of building drains used, depending on the type of foundation and kind of sewer you have.

The basement house with city sewer: The soil stack runs down into the basement floor. Under the floor, the building drain runs to the sewer. Sometimes there are inlets for a basement toilet and laundry. A cleanout is placed at the base of the soil stack. A floor drain is provided to drain the basement.

The basement or crawl space house with a septic tank: When a septic tank (or high-level city sewer) is used, the sewer connection usually enters the house above the floor of the crawl space or the basement, which requires a building drain suspended from the bottom of the floor joists. A cleanout is placed at the base of the soil stack. If a floor drain is used in the basement floor, it must be connected to the drain piping around the foundation. (See *Waterproofing the Foundation*.) If a basement toilet is desired, it will have to be a special type which is designed to flush upward to the building drain. Use the manufacturer's instructions to install this. A sump pump, which pumps up to the building drain, can be used to prevent water in the basement. Provide a 3-inch inlet for a toilet and a 1½-inch inlet for a sump pump.

The slab floor house: The building drain will have to be buried in the ground under the slab. Usually all inlets from the house, including the tubs, lavatories, toilets, sink, and laundry, must be run into the building drain. It is sometimes possible to connect the tub and lavatory to the soil stack by running the pipes through the wall, which eliminates the connections to the building drain. Connect the kitchen sink and laundry inlets in the same way, if possible. A cleanout must be provided. It can be located in the furnace or utility room, under the sink or lavatory cabinet, or in a closet.

The pier and beam foundation house: If the house is in a warm climate where freezing is not a problem, the plumbing can be hung from the bottom of the floor joists, and the building drain can run directly from the soil stack to the sewer. If the house has enough room under it in which to work, a cleanout will have to be placed in the house, as described for the slab floor house above. In cold climates, the sewer will have to be buried in the ground, as for the slab house. Be sure the pipe is buried as deep as the footings so it won't freeze, and insulate all the pipes running up to the house.

After you have determined which type of building drain you will use, you must choose the material. Check your local building code. If it refers to the National Plumbing Code, then you can use plastic pipe. If plastic pipe has been approved, by all means use it. It is strong and rigid, extremely resistant to chemicals and acids, and it fits easily into tight places. But most important of all, it is extremely easy to handle and work with. It is very light, and can be cut quickly with a hacksaw. The fittings are cemented to the pipe with a solvent-type cement which is painted on with a brush. Its cost is very reasonable, too. Plastic pipe for building drains is known as DWV (drainage-waste-vent), and comes in two types of plastic: ABS (acrilonitrile-butadiene-styrene) and PVC (polyvinylchloride).

If the building code has not been changed to include plastic pipe, check to see if an exception can be made. This is sometimes allowed when the code is soon to be updated. If not, the next material to choose is copper, which can be used only above ground. Almost all building codes allow copper, but check to be sure. Copper is very easy to work with: it's light and can be cut with a hacksaw. The fittings are soldered to the pipe, which is quite easy once you learn to do it.

The pipe must be clean and free from dents. Clean the end with steel wool until it is bright, and also clean the fittings if they need it. Buy a solder and flux mixture which is available in jars at a plumbing supply store. Paint this on the pipe and inside the fitting, then assemble the parts. All the openings in a fitting must be done at one time.

Heat the fitting evenly on all sides with a propane torch. Don't try to use an old-fashioned gasoline blowtorch because they are dangerous and too heavy. You can use two propane torches on the larger 3-inch fittings to heat them faster. When the flux starts to bubble out of the fitting, stop heating, and run solid core solder around the joint. This solder is about ⅛ inch in diameter, comes on spools, and is made especially for copper plumbing. The solder will be drawn into the joint by the cooling metal. Keep applying solder until the joint won't take any more, then go to the next joint. After all the joints are full, let the fitting cool before disturbing.

Copper is very expensive, but there is another material which is easy to work with and costs less: hubless cast-iron pipe. Its only disadvantages are that it is bulky and requires good supports. It comes in long straight pieces which can be cut with a hacksaw. Cut about ⅛ inch deep all around the pipe, then tap it with a hammer to break it. The pipe is joined to the hubless fittings with a slip-on neoprene gasket and a stainless steel clamp which is tightened with two screws. This pipe may be approved for underground use.

In case you have an antiquated building code which allows only hubbed cast-iron pipe, you should have a plumber install it. There are two types: leaded joints and rubber ferrule joints. Either one requires skills and tools which would not be practical for the do-it-yourselfer, and a poorly installed job can be very costly to correct. Some codes will specify hubbed cast iron for underground use only, and will allow copper, plastic, or hubless cast iron for above ground use. In this case, let a plumber put in the underground building drain and do the rest yourself.

After you have chosen the material, you can proceed with the installation. For slab houses, you must install the building drain before the floor is poured, so the exact location of the soil stack must be determined. The soil stack will be in the wall behind the toilet. If there are two adjoining toilets, the stack should be approximately equidistant from each. Locate the wall precisely from the floor plan. The exact position of the stack in the wall is not as important. Be sure that the stack will not be in the area where you want a medicine cabinet. The other outlets required for the toilet, bathtub, kitchen sink, and laundry must also be located. The bathtub drains are very difficult to get exactly right, so the best way is to locate them as nearly as possible and then leave about a one-square-foot opening in the floor slab. Connect the pipes after the house is framed—the hole in the floor is easy to patch.

If your house has a crawl space or pier foundation, you may want to put in the building drain before the floor is finished so you will have more room to work. Basement houses are usually completely framed before any plumbing is done. The basement floor is poured after the building drain and soil stack are completed.

If the building drain is underground, it is made of 3- or 4-inch pipe, depending on the material. This connects the soil stack to the sewer. If the house has more than one toilet, and they are spaced too far apart to share one soil stack, then two soil stacks are needed. A basement toilet should be near a soil stack, and uses a 3- or 4-inch pipe. All other branch drains which must run into the building drain are made of 1½- or 2-inch pipe. These are for tubs, lavatories, laundry, kitchen sink, and the basement floor drain. All corners in the building drain should be 45° bends. All branch drains should also connect at 45° by using Y's rather than tees. The entire building drain should slope down to the sewer at a slope of

¼ inch to a foot. A 10-foot length of pipe must therefore slope down a total of 2½ inches. If more than one soil stack is used, there must be a cleanout at the base of each stack. The basement floor drain should be 3 inches above the top of the footings, and is mounted on a P-shaped trap. When the house is framed and closed in, assemble the soil stack. Cut holes through the floor and the sole and top plates for the soil stack. Slide a 10-foot length of 3-inch pipe through the holes, and push the pipe up against the roof sheathing. Then mark and cut a hole through the roof. In case the soil stack interferes with a wall stud, just move the stud over and put another on the opposite side of the stack. If the soil stack interferes with the floor or ceiling joists, it should be moved if at all possible. If not, cut the joists and nail in headers. (See *Floor Framing*.) If the soil stack interferes with a roof rafter, use headers, or use two 45° elbows and go around the rafter. Use 45° elbows anywhere in the soil stack if it will ease construction. The soil stack will need tees for all the branch lines.

Any branch line more than 3½ feet from the soil stack will require a vent, so another tee is required above the fixtures, usually 2 feet above the floor, to accept the vent. The vents can be run through the attic instead of the wall if this is more convenient. All the tees are made to work in one direction only, and are also made to provide a drop of ¼ inch per foot. Be sure you install them correctly. The tees for the vents are installed inverted, so the gases will go up.

Next run the lines to the fixtures. We will discuss these one at a time.

Toilet: The toilet is connected to a closet floor flange on the bathroom floor. This flange is usually located 12½ inches from the finished wall, or 13 inches from the studs. Check the toilet dimensions to be sure this is correct for your toilet. This flange is connected to a 3-inch 90° elbow which runs to a tee in the soil stack. This tee can be either a 3 x 3 x 3 or a 3 x 3 x 3 x 1½, which has a 1½-inch side inlet for connecting to the bathtub. These tees come either right- or left-handed, or with two 1½-inch inlets, one of which can be plugged if not needed. If this fitting is not convenient, you can use a 3 x 3 x 3 for the toilet, and a separate 3 x 3 x 1½ for the tub. If you have two toilets back to back, use a 3 x 3 x 3 x 3 cross. The toilet must be as close to the soil stack as possible, and never more than 6 feet away. Maintain the ¼-inch drop per foot on the horizontal pipe. There are also wall-hung toilets which are costly, but excellent. The plumbing for these, usually copper, is special and is purchased with the toilet. Complete instructions come with the toilet.

Bathtubs: For slab houses, the building drain is brought up to a square-foot hole in the slab. The tub is now put in place over the hole. The building drain is connected to a 1½-inch trap, which is connected to the 1½-inch brass tub drain with an adapter. A swivel drum-type of drain trap makes this job easier. For other types of houses, the tub drain trap runs back to the soil stack. This is sometimes connected to the same tee as the toilet. If the tub is more than 3½ feet from the soil stack, use a vent line.

Lavatory: The 1½-inch drain is mounted in the wall behind and below the lavatory. It is about 16 inches above the floor. If the drain is 3½ feet or less from the soil stack, put a 3 x 3 x 1½ tee in the stack 16 inches from the floor and run a pipe across to the lavatory, then use an elbow to come out through the wall. The pipe should protrude about 2 inches from the wall. An adapter with a threaded nut is used to connect it to the 1¼-inch brass lavatory drain. The brass drain will be connected later. If the drain is more than 3½ feet away, use a 3 x 3 x 1½ tee under the floor. Run a 1½-inch pipe across to the lavatory, then use an elbow to come up to a tee 16 inches from the floor. The pipe will continue up to reconnect with the soil stack 2 feet or more above the floor to form the vent. If the stack is around a corner, it may be easier to go up into the attic to connect the vent. The drain can connect into the same tee as the toilet and tub if convenient.

Kitchen Sink: The sink can be connected to either the soil stack or the building drain. If the house has a slab floor, the sink connects directly to the building drain. If the house has a basement, you may want to run the sink directly down to the building drain using a 1½-inch secondary stack. This stack can also be used for a basement laundry. The stack must be vented either to the main soil stack, or up through the roof, whichever is easier. If you go through the roof, the last 2 feet should be increased to 3-inch size in freezing climates to prevent moisture from freezing over. If the sink is within 10 feet of the soil stack, the sink drain can run directly to the soil stack, but the pitch of ¼ inch to 1 foot is to be maintained. Run the vent as described above. Usually the sink is below a kitchen window, so the vent must run around the window.

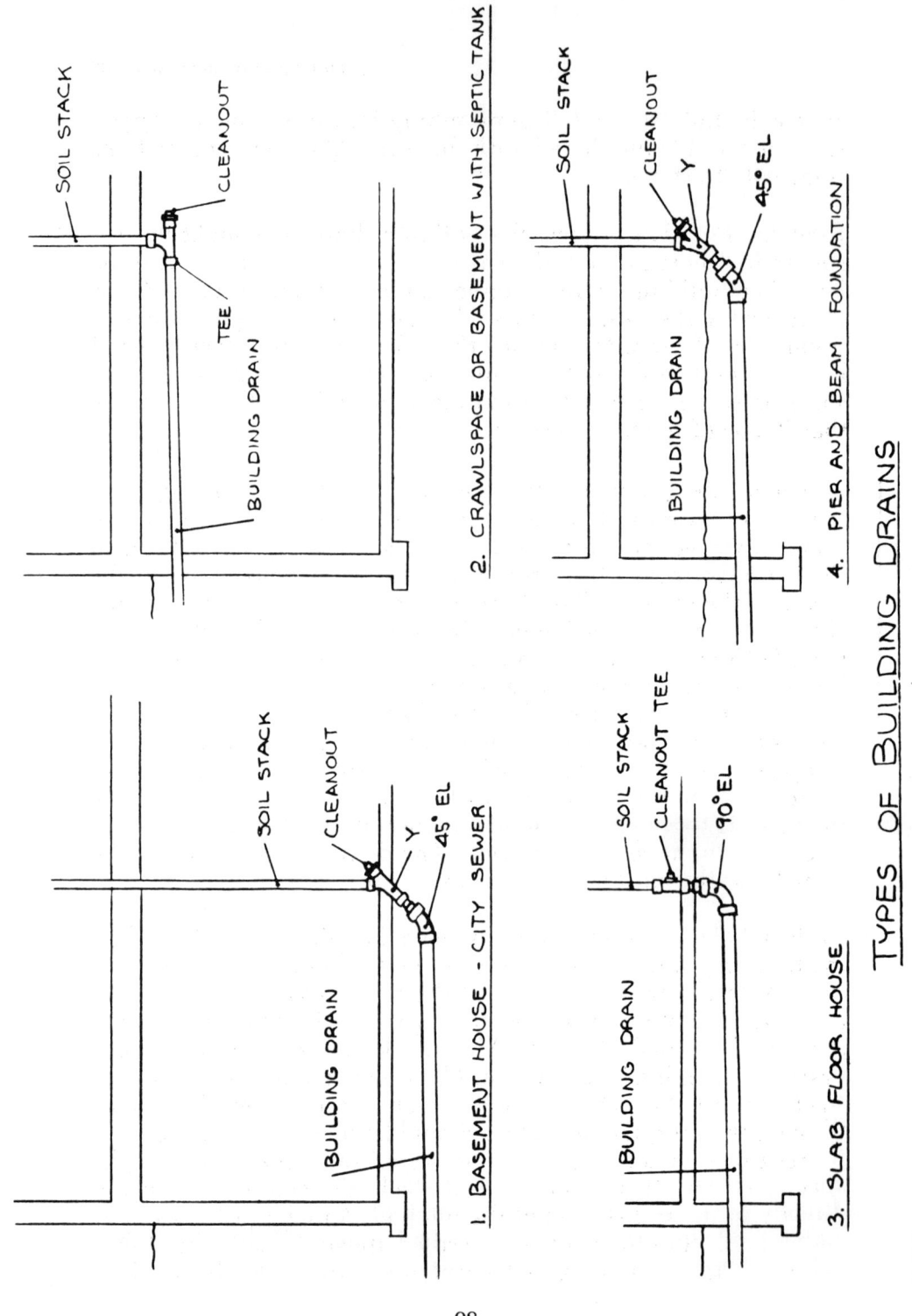

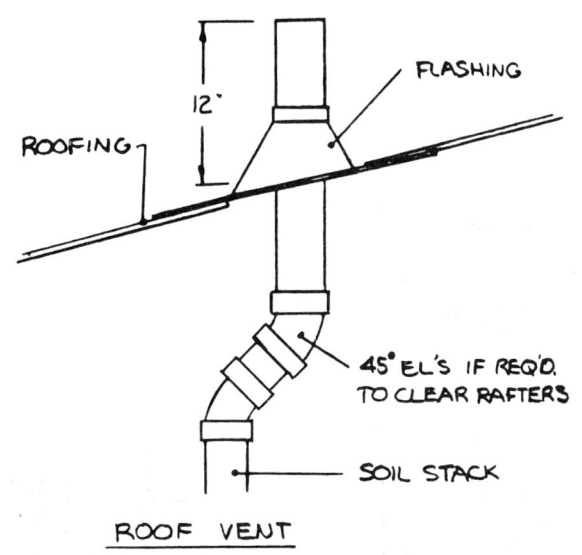

ROOF VENT

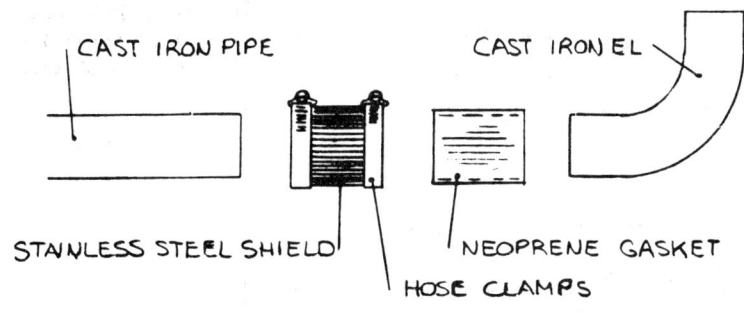

HUBLESS CAST IRON PIPE

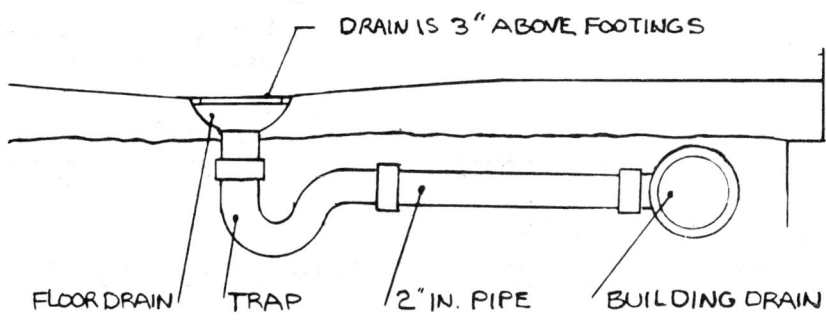

BASEMENT FLOOR DRAIN

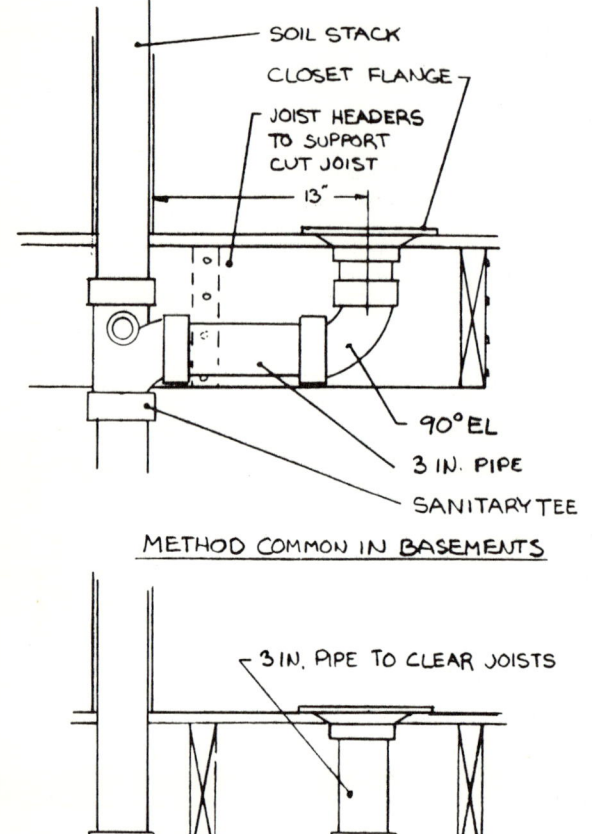

TWO METHODS OF CONNECTING TOILET DRAINS ACROSS JOISTS

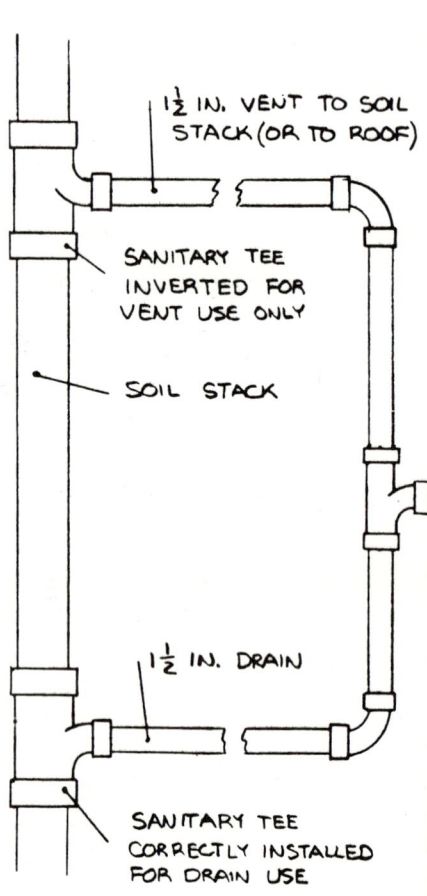

BRANCH DRAIN WITH VENT

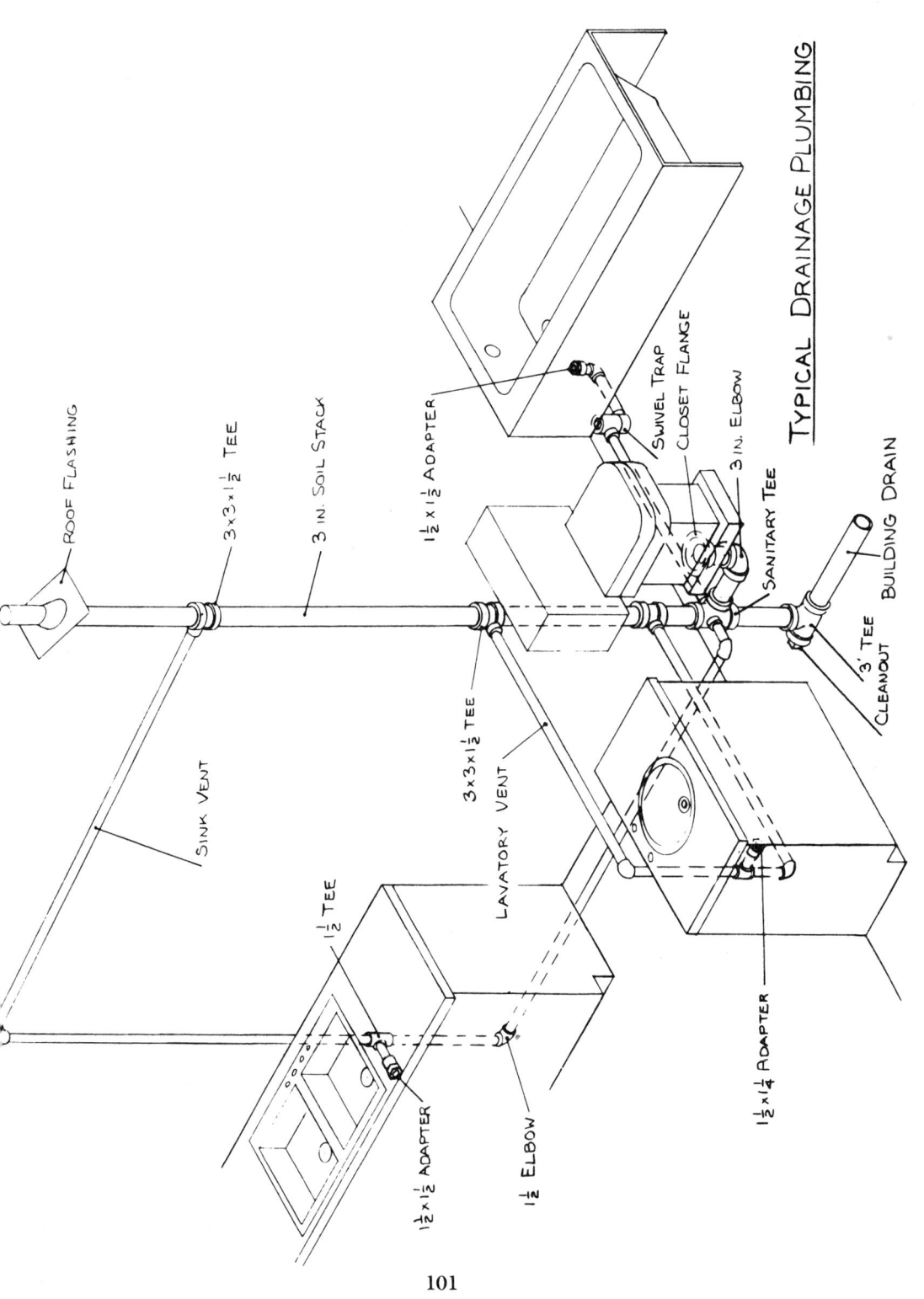

The sink drain should be insulated from the outside wall in freezing climates. Put insulation behind the pipe, but none in front, so the heat from the house can get to the pipe. The sink drain comes out through the wall about 16 inches above the floor and is connected to the 1½-inch brass sink drain with an adapter. The pipe should protrude from the wall about 2 inches. You may want to run the drain up through the floor rather than the wall, especially in very cold climates.

Clothes Washer—Dishwasher: The drains for these and other appliances should be planned and installed with the other plumbing. You will need a drain trap under the floor for either of these, but the exact location differs. If possible, get the instructions for the appliances you intend to use. If this is not possible, look at similar appliances at a store and get an idea how they are connected.

Water Softener–Central Air Conditioning: If these are close to the soil stack, they can be taken care of by a 3 x 3 x 1½ tee in the soil stack with either a P-type or a swivel-type trap, which are commonly used for connecting bathtubs. The hoses from the water softener and air conditioner will be inserted loosely into the trap.

After all the drain pipes are in, the roof flashing is put on the soil stacks. The flashing is slipped under the roofing and hammered tight to the soil stack. The joint is then coated with roofing cement.

The walls and floor can now be finished and the cabinets installed. The fixtures will be installed later. (See *Plumbing Fixtures*.)

24. Septic System

If you live in an area that is not served by city sewers, you will need to install a septic system. In some areas, septic tanks must be installed by an expert, but most areas will let an owner install his own, although a permit may be required. (See *Permits, Regulations, and Insurance.*) Take into consideration the location of any wells in the area. The septic tank must be at least 50 feet from any well and any part of the disposal field must be at least 100 feet from any well. It is a good idea to install the septic tank at the same time the house foundation is excavated since it requires a large hole.

Ready made septic tanks are available and are usually made of reinforced concrete or coated steel. Get the dimensions of the one you will use so you can prepare the excavation. The tank will be delivered by a special truck and placed in the hole. Check your local building codes for the size and materials required. If there is no code, the following is a rough guide to the size required. Two- or three-bedroom house–750 gallons; four-bedroom house–1000 gallons, but not less than a capacity of 100 gallons per person. If you will have an automatic washer or garbage disposal, you should allow an extra capacity of 25 gallons per person. Also, plan to use only biodegradable washing detergents.

Bury the septic tank at least 10 feet from the house. The removable cleanout cover must be between 8 and 16 inches below the finished grade. Figure this carefully by measuring from the bottom of the house foundation excavation. Remember that the cover will have to be removed each year to inspect or clean the septic system. This is usually done by a septic tank cleaning service. If you plan to clean the tank yourself, be sure it is located so that you can get a truck or trailer to it. The inlet of the septic tank must line up with a hole through the foundation. If the house has the bathrooms on the front side, the septic system can be installed in the front yard to keep the length of the building drain short.

The disposal field is laid after the septic tank is in place. The trenches are usually dug by hand because they are shallow, and the drop must be correct. The disposal field must not be closer than 10 feet to the house or property lines. The length of disposal line is determined by the absorption quality of the ground and the size of the septic tank it serves.

Dig a hole 12 inches by 12 inches and to the depth the disposal trenches will be. Fill the hole with water and keep it filled for at least 4 hours. Put a ruler into the hole, and pour in 6 inches of water. Record the time it takes for the water to seep away. The length of the trench is figured as follows:

Minutes Seepage	Feet of Disposal Trench		
	2 Bedrooms	3 Bedrooms	4 Bedrooms
0 to 12	85	130	170
12 to 18	100	150	200
18 to 24	115	170	230
24 to 30	125	185	250
30 to 60	165	250	330
60 to 90	190	285	380
90 to 180	250	375	500
180 to 270	300	450	600
270 to 360	330	500	660
over 360	Septic system cannot be used.		

There are two types of disposal fields used. For level ground, a loop, or closed, system is used; for sloping ground, a serial system is used. The serial system must be laid so that the parallel lines will be on fairly level ground, so the correct slope can be maintained. Dig all the trenches 24 inches wide. This is to give the correct absorption area. The trenches must all be at least 6 feet apart and no single section more than 100 feet long. The pipe will be laid at a slope of ¼ inch to each 6 feet. Cut a 2 x 4 twelve feet long. Nail a ½-inch-thick piece of wood to one end. Use this with a level to maintain the correct slope. Start at the outlet of the septic tank and dig the trench 6 inches deeper than the bottom of the pipe. Fill the trench with 6 inches of gravel and lay the pipe.

The pipe is 4-inch perforated plastic or fiber of the same type

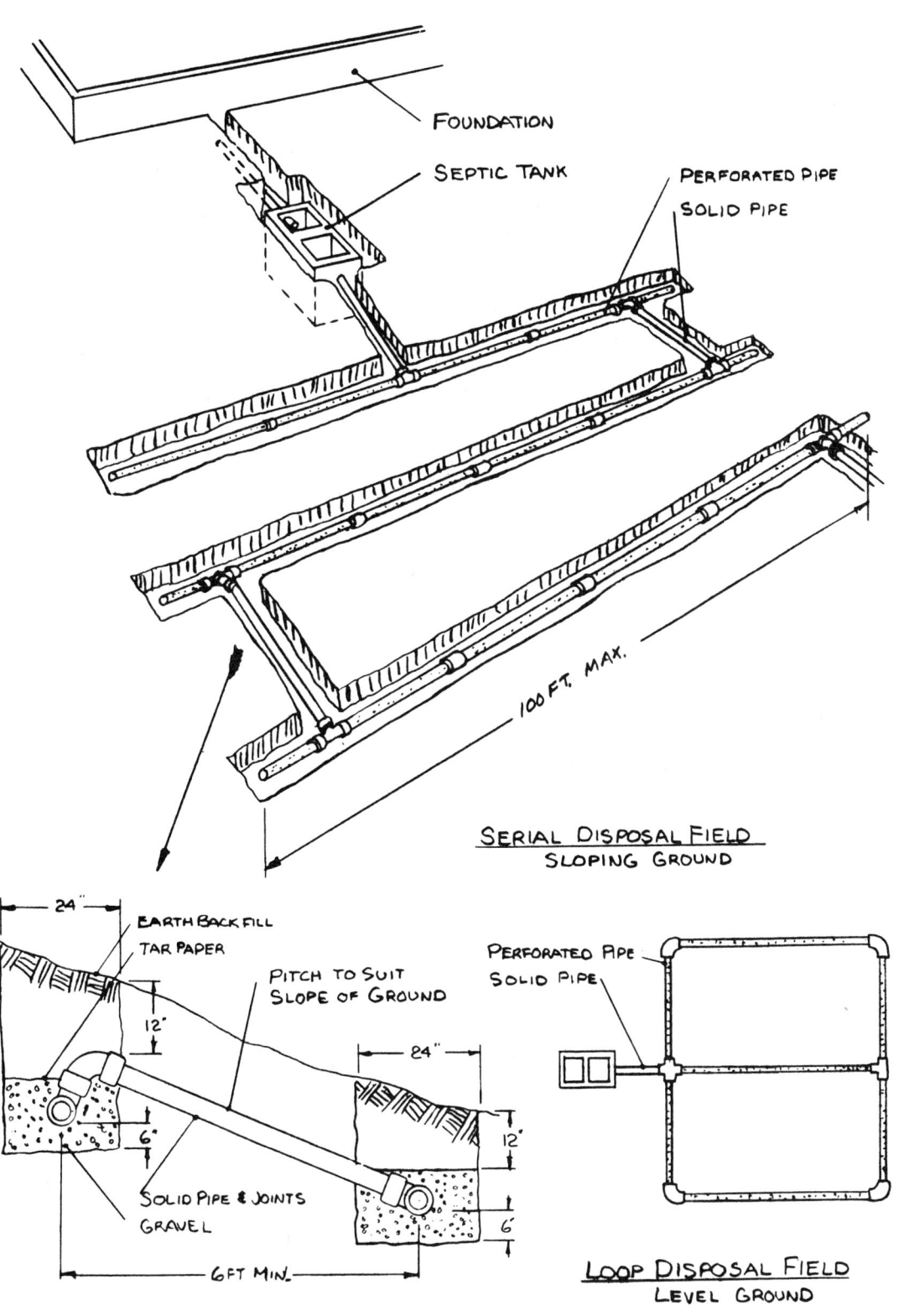

used for draining around basement footings. It is lightweight and comes in long lengths. The line from the house to the septic tank should be either steel sewer pipe or solid plastic. The line from the septic tank to the disposal field should be solid plastic or fiber. Cover the pipe with 2 inches of gravel and cover the gravel with tar paper. Backfill the trenches.

The following is an example of a septic tank system for an average house: A three-bedroom house for four people should have a 750-gallon septic tank. The disposal field is based on an average seepage rate of 6 inches in 30 minutes. From the chart, 30 minutes for a three-bedroom house will require at least 185 feet of disposal line. The field is divided into three lines each 65 feet long. This provides 195 feet of line, which is more than adequate.

More information on septic systems can be obtained by sending 35¢ to The Superintendent of Documents, U.S. Government Printing Office, Washington D.C. 20201. Ask for Public Health Service Publication No. 526.

If you are building in an area where it is very difficult or impossible to build a septic system, you can install a dry toilet which requires no water, sewer, or chemicals. It operates on the same principle as a gas incinerator. An electrical outlet, natural gas or bottled gas connection, and a chimney are required. These toilets are very good and are highly recommended by both sanitation and pollution experts.

25. Plumbing and Water Supply

After the house is framed, the water pipes can be installed. If the house has a basement, you may want to pour the concrete floor first. Ask your local water company what provisions should be made for the water meter. Some locales have underground meters or no meters at all.

The water pipes can be plastic, copper, or steel. Check your local building code. If it refers to the National Plumbing Code, then you can use plastic pipe, which is the best to use. The correct type is schedule 80 CPVC (chlorinated-polyvinyl-chloride). This is approved for 180° water. Cut the pipe with a hacksaw, and remove any burrs with a knife. The fittings are cemented to the pipe with a solvent-type cement which is painted on with a brush. The cost of plastic pipe is very reasonable. If the building code has not been changed to include plastic pipe, check to see if an exception can be made. This is sometimes allowed when the code is soon to be updated.

Copper is the next best material. It is cut with a hacksaw, and the fittings are soldered to the pipe. This is very easy to do. Clean the end of the pipe with steel wool. Buy a solder and flux mixture which is available in jars at the plumbing supply store. Paint this on the pipe and inside the fitting. Assemble all the openings in one fitting at the same time. Heat the fitting evenly with a propane torch. When the flux starts to bubble out of the fitting, stop heating, and run solid core solder around the joint. The solder will be drawn into the joint by the cooling metal.

If steel pipe must be used, consider doing the gas piping at the same time. Gas pipe is black steel which is basically the same as steel water pipe except that it is not galvanized. Use a stick-type pipe joint compound on the water pipe; the gas pipe takes a special type. A pipe-threading tool can be rented to thread steel pipe.

The following instructions apply to all but slab floor or pier and beam houses, which will be discussed later. Install a stop and waste valve in the water line where it enters the house. This valve has a small fitting on the side which can be used to drain the system. Use an adapter to connect to the ¾-inch plastic or copper water pipes.

To install the water pipes, first run a ¾-inch pipe from the water meter to the outside faucets and to the water heater. Usually a short length of pipe is used to bridge the space where the meter will later be installed. The local water company can give you information on this, or show you a similar installation. The ¾-inch pipe runs up to the floor joists and across the house to an outside faucet on each side of the house. If the climate requires, use freeze-proof faucets. These are about a foot long, so that the valve is well inside the house. Run another ¾-inch line to the water heater. The water heater must be located close to the chimney, unless it is electric. Be sure it is located correctly before the water pipes are connected. Run another ¾-inch line to the inlet of the water softener if you have one. All these lines are connected by using ¾-inch tees. Run ½-inch branch lines to the toilets. Use ¾ x ¾ x ½ tees for this.

The rest of the cold water piping is separate so that if a water softener is used the piping can be connected to the outlet. The piping is all ½ inch. It should be planned with the ½-inch hot water piping and installed at the same time. The hot water piping comes from the water heater. (See *Plumbing Fixtures*.) Run the hot and cold water pipes side by side, about 8 inches apart, to the kitchen sink, lavatories, bathtubs, and laundry. Where the pipes must cross, be sure they are far enough apart to prevent heat transfer. Be sure to run the hot water pipes to the left side of all fixtures. When doing the supply plumbing, keep in mind that the cold water pipe to the kitchen sink should be as close as possible to the water meter so you will have cold drinking water. Also, the hot water pipe to the tub and lavatory should be as close as possible to the water heater so you won't have to wait for hot water when bathing.

The lavatory supply pipes are run up inside the wall to a height of 20 inches from the floor and 8 inches apart and a tee is installed. A stub pipe protrudes 4 inches from the wall and is capped. Another pipe continues vertically for 12 inches and is capped—this is an air chamber to prevent "hammering" in the system. Install air chambers at each fixture. The cold water pipe for the toilets is run through the wall 8 inches from the floor, and 8 inches

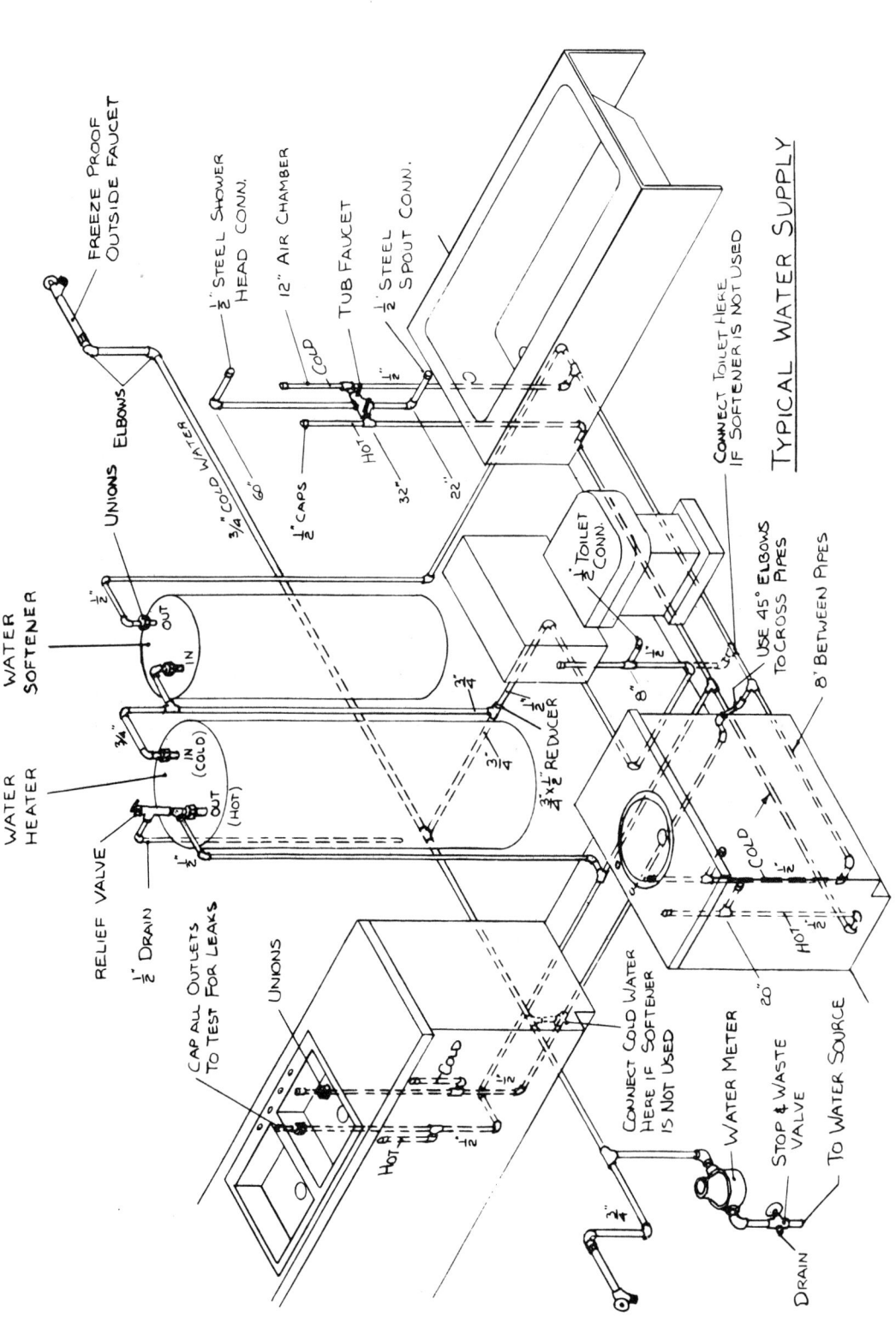

to the left of center. The kitchen sink can be connected through the wall or floor, whichever is most convenient. Laundry tubs are connected about the same as the kitchen sink, but other appliances require special connections. (See *Plumbing Fixtures*.) Don't install water pipes in exterior walls in cold climates.

The bathtub faucet must be installed while the rough-in plumbing is being done. It fits in the wall above the tub, 32 inches above the floor. Nail it to a 2 x 4 between the studs. Connect the water lines to the faucet with threaded adapters. Run a threaded steel pipe and elbow down to 22 inches from the floor, and screw in a nipple and cap for the spout. Run another threaded pipe and elbow up to 5 feet above the floor and screw in a 4-inch nipple and cap. This will be removed later to install the shower head. If you don't have a threading tool, measure and buy the pipe already threaded.

Slab floor houses or pier and beam houses in cold climates must be piped differently. The water supply pipe from the water meter is run up into the attic, and all the fixtures are piped from above.

To test the supply plumbing, all the pipes must be connected and capped. Connect a temporary pipe connection between the hot and cold water pipes for the water heater, if the heater is not already installed. Turn on the main shutoff valve and check for leaks. If you find a leak, you will have to drain the system and remove the leaking fitting. Reassemble with a new fitting, if required, and retest.

26. Well

Your water supply should be able to provide an adequate amount of drinkable water at a pressure of 30 pounds per square inch. The average house requires 75 gallons of water per day per person. A well is the most common water supply for areas which do not have city water; other sources are springs, ponds, and rain water cisterns. The well should be situated on a watertight platform, well above the surrounding surface. It should also be at least 100 feet from any septic system.

Wells are dug, drilled, or driven. Dug wells are 3 to 4 feet in diameter and up to 40 feet deep, and are rarely used anymore. Drilled wells are usually 2 to 6 inches in diameter and can go to depths of more than 500 feet. Drilled wells require expensive drilling equipment and skill and are costly to have done. Driven wells are more economical where they can be used. These are usually 1¼ to 2 inches in diameter and up to 50 feet deep. The flow rate of driven wells is low, but one or two gallons per minute is ample if a large storage tank is used.

Pumps are usually either submersible or jet type. The jet pump can be located away from the well and has a high capacity, but cannot be used with wells deeper than 250 feet. The submersible pump can be used at any depth. It is long and slim and is lowered down into the drilled well casing. When buying a pump, select a reputable dealer who can repair it quickly in an emergency. He will also be able to recommend the correct size and type of pump for your well.

Having a well drilled is a risky business, so find out who the best driller in your area is. Don't be concerned if he charges slightly more, since you must trust him to do the job correctly. Determine how deep the well will probably have to be by asking neighbors how deep they had to drill. Have the driller quote on the job and specify what size of well casing he will use and what equipment he will furnish. Larger well drilling concerns can supply the pump, storage tank, and pipe.

The well water must be stored in an area which is protected from freezing. The storage tank should have at least an 80-gallon capacity but the larger the tank is, the less often the pump will have to start and stop. If the tank will be in the house, it should be an insulated model to prevent condensation.

If the well water is too hard to use for washing, you may need a water softener. This can be installed in the plumbing system.

27. Plumbing Fixtures

The plumbing fixtures must be chosen to fit your budget since they come in a wide range of prices. The water heater can be either gas or electric. If you have gas heat, it is a good idea to have a gas water heater also. The following chart is a guide to the correct size water heater.

Size of family	Gallons	BTU Input per hour
2 or 3 people	30	40,000
	40	35,000
4 or 5 people	40	75,000
	50	50,000
6 or more people	50	75,000

Gas water heaters will have to be located close to the furnace so that they can be connected to the chimney. Be sure to get a relief valve to use with the water heater. Use adapters to connect the water heater to the water pipes. Use a ¾-inch tee to provide a side connection for the ½-inch hot water pipe; the relief valve goes on top of this tee. The cold water pipe should be ¾-inch.

Buy bathtubs, toilets, and lavatories from one source so they will match. The tub can be either steel or cast iron, although steel tubs usually come only in white. Cast-iron tubs are more expensive but more resistant to the enamel chipping. They come in colors, but many new homes are using white with brightly colored ceramic tile and accessories.

Toilets are available at many prices. The higher-priced models all work about the same, so appearance is the main difference. Wall-hung models will require special plumbing consideration, but are very nice, especially if you plan to have carpeting in the bath. Instructions for doing the plumbing come with wall-hung toilets. A wax or plastic ring is used to connect floor-type toilets to the closet floor flange. If the toilet is installed before the finished floor, it can be reset later. Use a chromeplated toilet-to-wall supply pipe with a valve to connect the toilet to the water supply.

There are several styles of vanity lavatory bowls. One type mounts in the cabinet top with a metal rim. Another mounts under a special top with a finished hole for the lavatory. A third type is a china top which covers the entire vanity. The faucets and drains are also available in many types; the drains are almost always mechanical. Some faucets do not have washers and are guaranteed leak-free for ten years. Use an adapter to connect the ½-inch water pipes to the ⅜-inch pipes on the faucet; no shutoff valves are needed. The drain is connected with a 1-¼-inch wall-connection drain trap.

There are two types of kitchen sinks. One type has a built-in rim and is usually stainless steel. The other kind—usually enameled and available in colors—is set into the counter top with a separate rim. Almost all kitchen sinks are double-bowl sinks. The faucets are often the single-lever type and have a spray head on a flexible hose. Use adapters to connect the ½-inch water pipes to the female pipe connection on the faucets, and put a union in each pipe to allow the faucet to be removed. Use basket strainers for the sink drains and connect to the house drain with a floor- or wall-connection sink drain trap and connecting waste.

Other fixtures require similar connections. These are laundry tubs, water softeners, automatic washers and dishwashers, and they usually come with very good instructions on how to install the plumbing. If you are planning to use any of them, ask the dealer to let you see the instructions before you buy. You may even be able to make a copy of them so you can prepare the plumbing in advance.

28. Plumbing for Gas

Ask your local gas company for requirements and pipe sizes for connecting the gas meter. In general, if your furnace is within 30 feet of the gas meter, you can use 1-inch gas pipe for furnaces up to 230,000 BTU. Use black steel pipe and black iron fittings. Also, use special thread compound made for gas pipe, and apply it sparingly to the male threads only so none will get inside the pipe.

The gas company will install the meter. Starting from the meter, run the 1-inch pipe to the furnace. If the house has a basement, the pipe will be hung from the basement ceiling, otherwise it will be in the crawl space or attic. Run the 1-inch pipe across the ceiling to the furnace and install a tee. Let the vertical pipe run down to the floor with a cap on the bottom. This is called a drip leg and is intended to catch the dirt and moisture. Install another tee horizontally in line with the gas valve in the furnace. Reduce to the pipe size of the gas valve, usually ½ inch, install a union and shut-off valve, and connect to the furnace. Use ¾-inch pipe to go to the water heater and install a ½-inch shutoff valve and union at the heater. Use a drip leg wherever a vertical pipe is used. Run another ¾-inch pipe to the other gas connections and install ½-inch shutoff valves at each outlet. Appliances such as gas clothes dryers and kitchen ranges are connected with brass flexible tubing to allow them to be moved for cleaning and servicing. Connect the brass tubing to the shutoff valve.

If you don't have pipe-threading equipment, or don't want to thread the pipe yourself, you can measure for each piece and do one at a time, buying the pipe already threaded from a plumbing supplier. Use nipples of various lengths to connect to the appliances to take up any error.

When the piping is installed and the appliances are connected, close all the valves and turn the gas on at the meter. Make a soapy mixture of detergent and water and put it on every pipe joint and valve to detect leaks. If bubbles form, correct the leak immediately. Don't use a match: it is not only dangerous, but it won't detect leaks too small to affect the flame.

Bottled gas, generally LP gas, is sometimes connected by using steel tubing. Your LP gas dealer can recommend the size and type you will need, and he'll also have the gas supply tank you will need.

29. General Heating

There are three types of heating systems, all of which are easily installed. Electric heat, which is very easy to install if baseboard units are used, has some advantages, such as independent temperature controls for each room. It warms up very quickly and is completely quiet. A special 200-amp. electrical service is required. (See *General Electrical Service*.) One disadvantage is that there is no ductwork and therefore central air conditioning cannot be added to the system.

Hot water baseboard heat also offers easy installation. The baseboard units are connected with copper pipe, and the furnace itself is very small and quite simple to put in. The main advantages are quiet operation and very even heat. The disadvantages are that it is slow to warm up or cool down, and there is no ductwork to use for air conditioning. Fuel is either oil or gas.

The most common heating system is forced warm air, which is excellent for a number of reasons. The temperature is easily controlled with a single thermostat and with adjustable baseboard outlets. It warms up very quickly and cools quickly. The installation cost is lower than hot water, and two previous disadvantages have been overcome. The new blowers are so quiet that they are barely perceptible. The filter systems are so good that the house will actually stay cleaner than with hot water or electric heating systems, which recirculate dirt and dust by convection. With adequate outlets, the air movement is very slow and this results in even heat. The main advantage to warm air heat is the possibility of adding air conditioning, humidifiers, and electronic air purification units. The fuel used is usually either oil or gas, but electric warm air furnaces are also available. Forced warm air heat is recommended for amateur builders.

The type of fuel to be used is determined by availability and cost. Gas is the most common and usually lowest in cost, although in some areas electric heat is competitively priced. Check the availability of both because sometimes there are waiting lists. If gas is not available and electricity is too expensive, you can use oil or bottled gas. The notion that oil heat is dirty is completely false. The furnace works exactly the same as with gas. There may be a slight oil odor if the furnace is in the house, but it is certainly not objectionable. An oil tank is usually buried in the yard in an area which can be reached by the oil truck. Bottled gas tanks are usually placed above ground and are also refilled from a truck, or exchanged for full units.

The easiest way to purchase a good heating system is to take your floor plans to heating dealers and get bids on all the required materials. They will figure the correct size furnace and give you a list of what they will provide for a certain cost. Electric and hot water heat must be figured accurately in order to get the baseboard units the correct length. Local dealers are familiar with the climate in your area and will know how to size the furnace for temperature and wind conditions. Sears and Wards will also sell packaged heating systems complete with instructions. These systems usually make use of prefabricated chimneys, so have this included with the system. Also, if you have forced warm air heat, have the ductwork designed for air conditioning if there is even a possibility that you will some day use it. If you have a basement, install extended plenums rather than round pipes for a neater appearance.

30. Hot Water Heating

Baseboard hot water heat is available in various arrangements. There are two basic types of systems: the one-pipe system and the series-loop system. The one-pipe system has all the baseboards connected in one loop so no individual control is possible. This sometimes causes some rooms to be too hot or cold, but is usually adequate and economical. The series-loop system has independent baseboards for each room, but costs more and is harder to install. Each baseboard has a control valve to balance the system. Either system is controlled by a single thermostat. By splitting the house into two or three zones and using a separate loop for each, individual zone thermostats can be used. This costs even more. When getting a quotation, know which system is being quoted. Different types of baseboard units are also available, but the most common is the convection type made of a copper tube with aluminum fins and a steel cover. If you have a one-pipe system, make sure the baseboard covers have adjustable dampers so you can control the heat in each room.

   Installing the system is fairly easy. Sears and Wards provide full installation instructions with their systems. The baseboard units are nailed to the finished walls, and holes are drilled in the floor to allow the pipes to go down into the basement or crawl space. The pipes are then soldered together as indicated by the heating dealer's plans. The furnace is located in the basement or furnace closet, and has a small electric pump and an expansion tank which are connected to the system. The system is connected to the water supply to keep the lines full. A thermostat is used to turn the furnace on or off and the electric pump is automatically turned on when the water is hot.

Prefabricated Chimney

Use a prefabricated chimney, with an inlet for a gas water heater if required. This is easily installed by cutting a hole through the floor, ceiling, and roof and slipping the section through. The sections lock together. Firestop spacers are provided to keep the pipe away from the surrounding framing. The location for the chimney is shown on the plans, and a prefabricated housing is installed on the roof. This must be cut to the correct slope of the roof, and must extend 2 feet above the highest part of the roof.

31. Forced Warm Air Heating

Forced warm air heating systems are available in various types, depending on the house and climate. The most common is the basement furnace with extended plenum ducts, which are rectangular-shaped pipes running the length of the house, one for warm air and one for return air. Round pipes are used to connect the extended plenum to the warm air baseboard outlets and are located between the floor joists. This makes a neat basement. The return air ducts are simply closed joist spaces connected to the return air inlets, which are located in the spaces between the wall studs.

Houses with crawl spaces can have the same type of system as above or can simply have round pipes going directly from the furnace to the inlets and outlets. These pipes hang below the joists and cost less but would reduce head room if used in a basement. The furnace can be a horizontal unit hanging in the crawl space, or a counterflow unit located in the house with a hole cut through the floor. The counterflow furnace is much easier to work on.

Houses on slabs usually do not have baseboard outlets. The warm air pipes are installed in the attic using either an extended plenum or individual pipes, and the warm air outlets are located in the ceiling next to the outside walls. The furnace can be either an up-flow type with a return air grille at the bottom, or a horizontal unit in the attic with a single large return air duct extending down to the floor level. A much better system is possible with slab floors by installing the ducts in the concrete slab and using baseboard outlets. (See *Heating: Perimeter for Slabs.*)

Install the forced warm air heating system before the walls are finished. If the house has a basement, you will want to finish the floor first. Start with the furnace, and locate it near the chimney so that the pipes will be in the best location. For basements, the

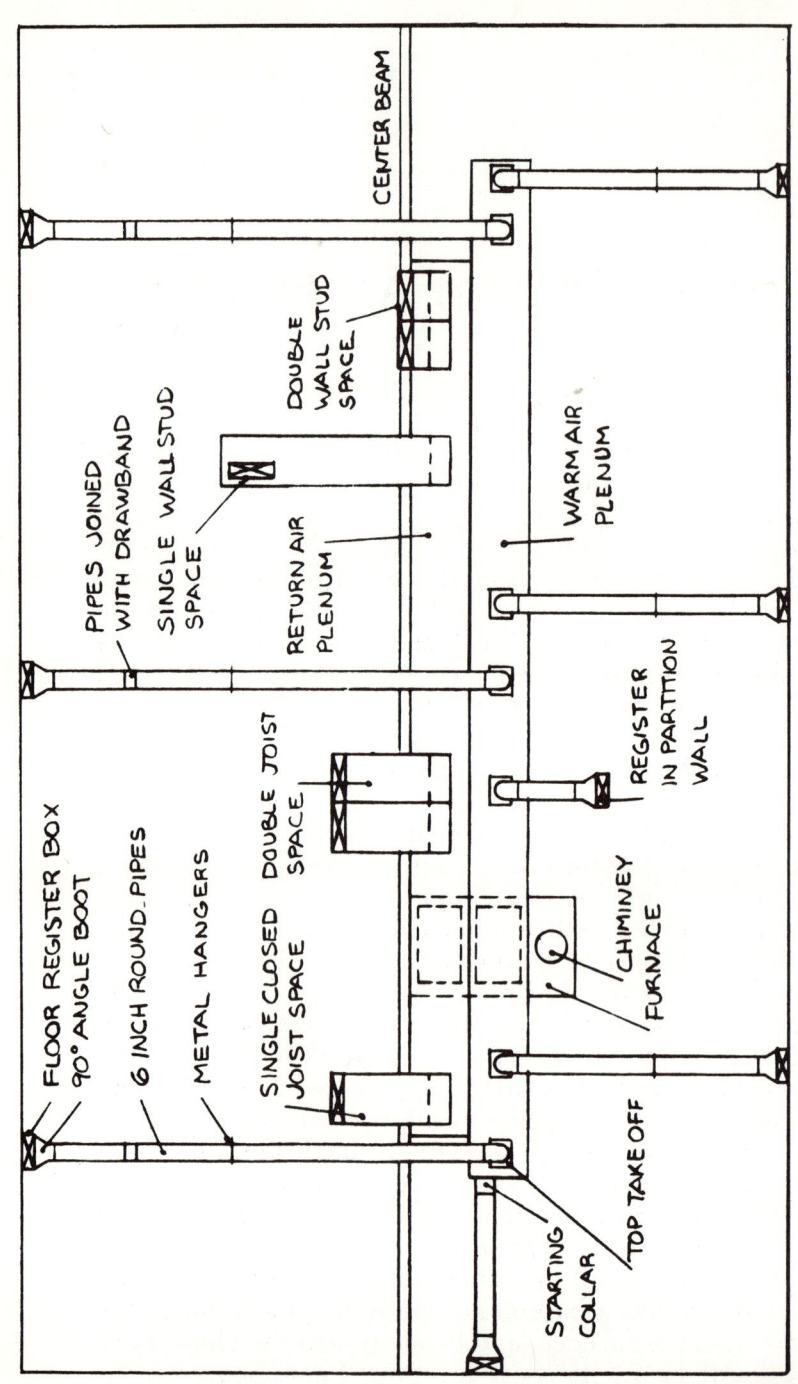

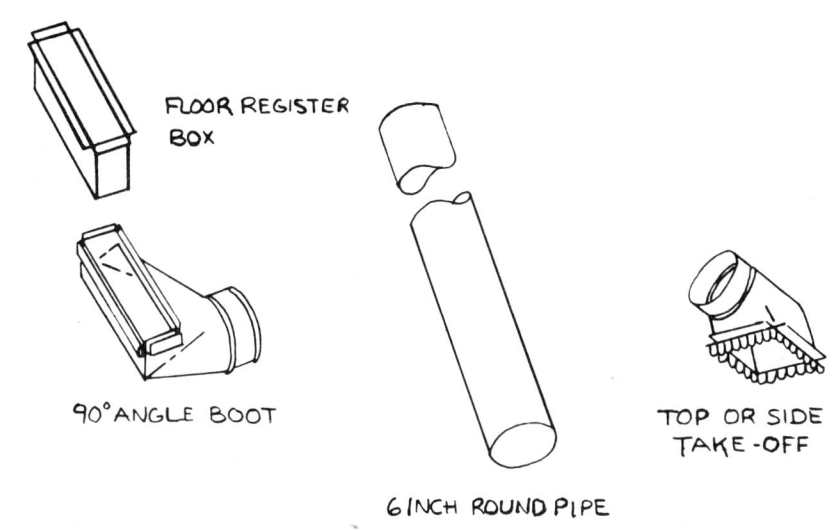
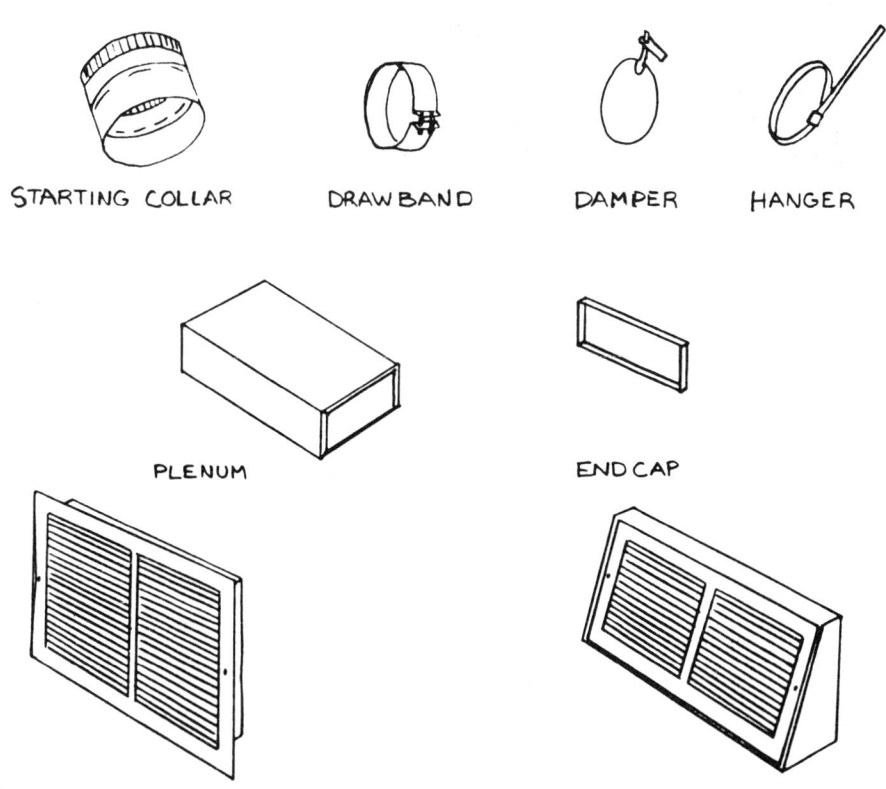

FORCED WARM AIR HEATING PARTS

return air plenum should be right up against the center beam, and the warm air plenum should be next to it. Have the plenums extend even with the farthest outlet on each end of the house, if possible, for the best appearance.

The return air plenum has rectangular holes cut in its top side to connect to the floor joists below where each inlet will be. The joists are closed in with sheet metal. A hole is cut through the sole plate and the subfloor to connect the joist space to the space between the studs. Nail a 2 x 4 block between the studs just above where the inlet will go.

The warm air plenum is hung from the joists with metal hangers. The round pipes are connected to the top of the plenum with a fitting called a top or side take-off, and to the floor register box with a fitting called a 90° angle-boot. The baseboard outlet will be installed later when the interior trim is put in. Cut a hole through the subfloor to fit the floor register box. The round pipe is held up with special hangers, and the lengths of pipe are connected by slipping the ends together. To connect the last piece, a drawband is fastened around the joint; all joints are fastened with sheet metal screws. Each round pipe has a damper in it.

32. Heating: Perimeter for Slabs

To install ducts in a concrete slab, the location of the furnace and all outlets must be determined before the slab is poured. Mark the positions of these, and dig a shallow trench all around the inside of the foundation. Dig a hole under the point where the furnace will sit, and pour a small concrete slab 12 inches below the level of the house slab. Build a box the size of the furnace plenum out of 4-inch concrete blocks or bricks, with outlets made of 6-inch clay sewer pipe. Your furnace supplier will determine the correct number of outlets. These sewer pipes run out to the perimeter of the house. Cut the sewer pipe by marking all around with a cold chisel, then striking it with a hammer. Use 45° elbows as required to get the pipes evenly spaced about the perimeter. Use tees to connect the pipes to another pipe running all around the perimeter of the house; use 90° elbows in the corners. All this pipe must be of the type which has waterproof hubbed joints. The perimeter pipe should be 2 inches below the finished floor surface, and the trench under the pipe filled with gravel. At the location of each outlet, wire a board to the top of the tile. Make the boards ⅛ inch bigger than the outside measurements of the floor outlet and 2 inches thick.

After the slab is poured, locate the outlets and remove the boards before the slab is completely hard. When you are ready to install the outlets, use a cold chisel to cut through the tile.

33. Air Conditioning

When installing the heating system, consider installing central air conditioning at the same time. Central air conditioning is far more effective than window units, is silent, and will usually cost less than $500. The easiest way is to use a forced warm air heating system and put the air conditioner evaporator coil in the hot air duct of the furnace. From some manufacturers you can buy the furnace and air conditioner preassembled. Have your local dealer select the correct size unit and give you a quotation on the system, but get more than one quotation and compare the sizes and prices. The only proper measurement of air conditioner size is in BTU's, not tons or horsepower. Remember that too small a unit will not cool the house properly, and too large a unit will not dehumidify properly. This will give you a cool but damp house. Sears and Wards are excellent sources for central air conditioning.

Installing the air conditioner is very easy. First, be sure to purchase precharged components and tubing. If you don't, you will have to pay a refrigerator repairman to charge the system. Locate the condenser unit outside the house as close to the furnace as possible but away from bedroom windows and away from your nearby neighbors' windows, as there is some noise outside the house. Pour a small concrete slab on which to place the condenser, which should be 24 inches from the house to provide proper air flow. Cut a hole in the wall for the refrigerant and electric lines. Install the evaporator in the furnace plenum (the instructions for this will come with the air conditioner). Connect the refrigerant lines from the condenser to the evaporator.

Provide a 230-volt circuit for the condenser. (See *General Electrical Service*.) Use a heating–air conditioning thermostat. All wiring instructions are provided by the furnace and air conditioner manufacturers, and if you buy a preassembled furnace and air conditioner, most of the wiring is already done.

If you don't have forced warm air heat, you can install air conditioning in the attic and provide ductwork for it. Get quotes on the complete system from your dealer. The cost will be higher because of the ductwork and the large blower that is needed in place of the furnace fan.

The chances are that unless you live in an ideal climate you will be installing central air conditioning sometime in the future. In case you are not planning to put in air conditioning when you build your house, consider installing the ductwork and thermostat required for it. This will eliminate the very high cost of installing these after the house is finished.

34. General Electrical Service

Installation of the electrical service for a home is not difficult, but it requires an understanding of the principles involved. Some people are afraid to work with electricity or feel that it is too complicated to understand, but anyone who has the ability to build a home can quickly and easily learn all that is required to safely wire a house.

The following terms are used in electrical work and must be understood.

Circuit: One complete loop of wire—the path of the current—from the fuse box to the outlets or an appliance and back to the fuse box. Separate circuits are used for large or special appliances.

Ampere: The unit for measuring electric current of flow, usually abbreviated "amp." This term is used in sizing wires or fuses, both of which must be larger if more electricity is to flow through them. Most house electrical circuits are 15 amps. Larger appliances require more electrical flow, so 20- to 50-amp. circuits are also used.

Volt: The unit of electrical pressure. Most house electrical circuits are 115 volts. Appliances which use larger quantities of electricity, such as air conditioners, electric ranges, or clothes dryers, would require very large wires. To reduce the wire sizes, higher electrical pressure, or voltage, is used to push the power through the smaller wires. These large appliances use double the standard house voltage, or 230 volts. Door bells or chimes require very little electricity, so they are operated on a special circuit of only 6 to 18 volts.

Watt: A unit of power—the measurement of electrical quantity. Amperes multiplied by volts equal watts. All appliances and light bulbs are rated in watts, and this number is the quantity of electricity they use. A large appliance which is rated at 2300 watts could take either 230 volts times 10 amps. or 115 volts times 20 amps. Since 20 amps. is more than the capacity of the usual 115-volt household electrical circuit, a 2300-watt appliance would be on a special 230-volt circuit.

Watt-hour or Kilowatt-hour: The measurement of quantity of electricity used for a certain interval of time. Electric meters measure the power used for a month, or some other period, in kilowatt-hours. This is the figure on which your electric bill is based.

There are other electrical terms which are not required for electrical installation but with which you should be familiar. "AC" stands for alternating current, which is the type of electricity used in all houses in the United States. "DC" stands for direct current which is sometimes used in factories and some foreign countries.

"Sixty-cycle current" refers to current which alternates 60 times per second, and is the type used in the United States. Some foreign countries have 50-cycle current. "Single-phase current" is another term which applies to common household electric current. Three-phase current is used in factories or farms, where very large electric motors are used. "HP" stands for horsepower; one electrical HP equals 746 watts.

When you wire your own house, you are usually required to get a permit and pass an inspection by the electrical inspector before the wiring is covered. (Read the chapter on *Permits, Regulations, and Insurance.*) Be sure you understand the local code if there is one. Most local codes refer to the National Electrical Code, but if your locality has its own code, it may differ from the methods described below.

The first step in planning the electrical wiring is to determine what amperage will be required. Most houses have 100-amp. service. If a great many large appliances will be used, such as a range, clothes dryer, water heater, and air conditioner, a 200-amp. service is recommended. An easy way to check if 100-amp. service is adequate is to multiply the square feet of living area times three watts and add 3000 watts for the kitchen, 1500 watts for the laundry, and the actual wattage from the name plates on the range, water heater and air conditioner. If the total comes to more than 23,000 watts, you may want to consider 200-amp. service, especially if you plan to use a lot of other appliances. If electric heat is used, the service usually must be 200 amps. Check this out with an electrical heating contractor or supplier. In some areas with large homes, 200-amp. service is required by regulation.

Next determine how many 115-volt electrical circuits are required. You must have one for every 500 square feet of floor space plus two for the kitchen, one for the laundry area, one for the furnace, and one spare. Also, if you have a workshop, add one circuit for it. If you plan to use 115-volt window air conditioners, add a circuit for each air conditioner. Now determine how many 230-volt circuits are required. You will need one for any of the following: electric water heater, electric range, electric clothes dryer, well pump, central air conditioning unit or 230-volt window air conditioner, and one spare.

The house will be wired with one of three systems: conduit, armored cable, and nonmetallic-sheathed cable.

If your local building code calls for conduit, the conduit is placed in the walls by notching the studs. Conduit connectors are used to connect the conduit to the boxes, and conduit couplers are used to connect sections of conduit. The conduit is easily bent with a conduit bending tool, but it is difficult to install and requires considerable planning to keep the amount of conduit to a minimum.

Some codes require flexible armored cable sometimes called BX. This is much easier to use than conduit and will allow you to wire in difficult areas with ease. Special connectors are used to connect to the boxes.

Most modern codes specify nonmetallic-sheathed cable sometimes called Romex, which is also very easy to use and costs about half as much as armored cable. It is waterproof, and some types can even be buried underground or in masonry. Boxes are available with built-in clamps for nonmetallic cable. If possible, use nonmetallic cable for the house and conduit for any exposed wiring in the garage or basement. Use armored cable for difficult areas of exposed wiring.

Study the floor plan and note the location of the receptacles and switches. Every room should have a wall switch wired to at least one receptacle or to a ceiling light. Ceiling lights are used in most rooms, except living and bedrooms, which usually use table lamps plugged into receptacles controlled by a light switch. Closets other than the walk-in type often do not have lights. Receptacles should be located in all rooms so that there is no more than 12 feet between receptacles, measured along the wall.

The following electrical symbols are used on floor plans:

  Ceiling or wall light fixture

  Duplex receptacle

S   Single-pole wall switch

S³ Three-way switch to control a receptacle from two points

S⁴ Four-way wall switch to control a receptacle from three or more points

――― Wire connecting switches and the receptacles to be controlled

Large rooms with two or more doors sometimes have switches at each door. Plan to have switches located so that you can go from one end of the house to the other at night without having to be in the dark. If you have a garage, the garage lights should have switches at the house door and at the garage door. You may also want a switch at the rear door. The front exterior lights should be controlled from the garage or front door. A special switch with a timer can be used to keep the lights on for a few minutes after you leave the house. This switch is installed just like an ordinary switch.

The duplex receptacles in the living room and bedrooms can be split-wired so that the top receptacle will be controlled by the switch, and the bottom receptacle will always be on. The living-room and bedroom receptacles are sometimes equipped with a dimmer control which will vary the lights from a soft glow to full brightness. This switch is also installed just like an ordinary switch.

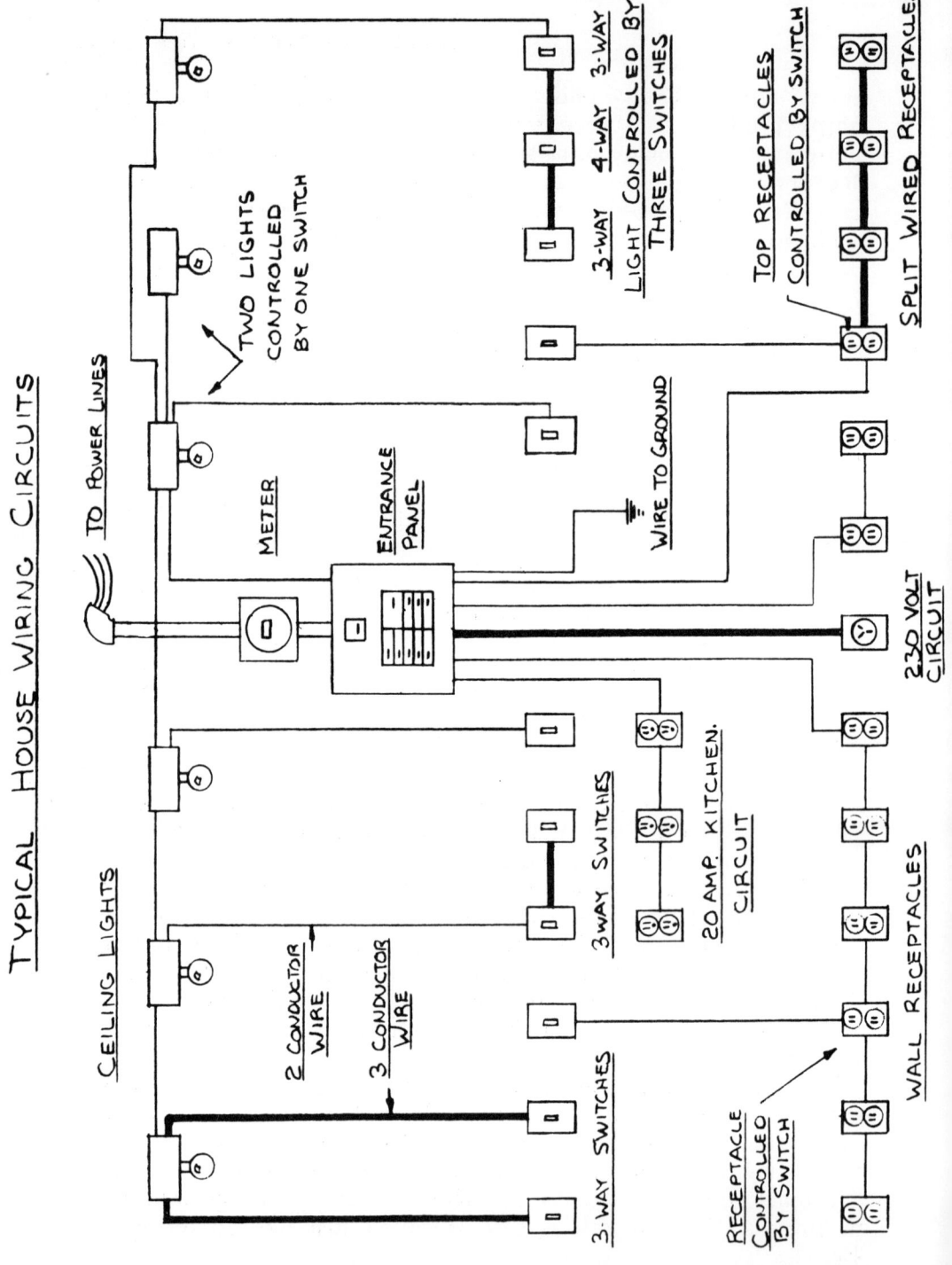

It is important to remember all the devices and appliances which require wiring, so the following list is provided as a guide to outlets you may need. Appliances which sometimes take 230 volts are noted.

Ceiling lights	Bathroom fan	Sump pump
Crawl space light	Automatic washer	Door bells or chimes
Yard lights	Electrical heat (230V)	Garage receptacles
Garage light	Water heater (230V)	Dishwasher (230V)
Attic light	Intercom	Freezer (230V)
Wall receptacle	Shop power tools (230V)	Bathroom heater
Basement lights	Wall clock receptacle	Attic fan
Porch and patio lights	Outdoor receptacles	Furnace fan or pump
Bathroom receptacle	Refrigerator	Well pump (230V)
Counter top receptacles	Cooktop (230V)	Swimming pool equipment
Basement receptacles	Kitchen fan	
Garbage disposal	Automatic dryer (230V)	Garage door opener
Oven (230V)	Air conditioning (230V)	

35. Electrical Installation

Go to the local power company and get any information they have about connecting the service entrance, which is the first thing to install. If your area has overhead wires, the power company will run three large wires to your house. You will need to supply a metal hook to support these wires, a ground connection, a meter socket for the power company's meter, and a circuit breaker entrance panel.

Locate the service entrance at a point which is close to the utility pole, and directly above a point where the circuit breaker entrance panel can be placed. This panel is usually located in the basement, garage or utility room, or in an inconspicuous area of the house, such as in a closet. It must be on the outside wall, or close to it, so that the large service pipe can be easily connected.

If the roof of the house is low, the service mast must extend two feet above the roof. If the roof is high, the mast can stop on the exterior wall 12 feet above the ground. Also, be careful that the location does not detract from the appearance of the house. If your area has underground wiring, the power company will give you complete details on how to provide the service entrance.

Assuming your house has a low roof, hold the conduit under the soffit at the point where the meter will be located, and trace the outline on the soffit. Make a hole with a drill or keyhole saw and slip the pipe up through it until it touches the roof sheathing. Have someone hold it up. Tap on the roof from above until you locate the conduit, then drill a hole. Enlarge it to let the conduit slip through. If you have 200-amp. service, use 2-inch conduit and fasten an eccentric fitting to the bottom of it; then connect it to the meter socket with a 1¼-inch close nipple. If you have 100-amp. service, use threaded 1-¼-inch conduit and screw it directly into the meter socket. Align the conduit and meter socket, and fasten the conduit to the wall with conduit supports. Screw the meter socket to the wall 5 feet above the finished grade. Slip the roof flashing over the conduit (see *Flashings, Vents, and Gutters*) and screw on the service head. A metal hook or insulator is fastened to the roof joists, or if you are not going through the roof, to the wall studs. This supports the wires for the utility pole. Come out of the bottom of the meter socket with 1-¼-inch conduit.

If you have a basement, drill a hole through the joist header and pass a short length of conduit through it. Connect an entrance ell to both ends and connect the other ends to the meter socket and the service entrance panel. Mount the service panel on the basement wall at eye height. If you have no basement, drill through the wall directly below the meter socket and use an entrance ell to go from the bottom of the meter socket through the wall into the back of the service entrance panel. Mount the panel flush in the wall, at eye level.

The meter socket is sometimes supplied by the power company; any local electric store can help you select the other parts you will need. Or, try to use parts similar to other houses in your area.

Purchase a series-wired circuit breaker entrance panel. Add up the circuit breakers you require. The 230-volt circuits require a double circuit breaker, so if you need six 115-volt circuits and three 230-volt circuits, you will need a panel with twelve circuit breakers and a 100- or 200-amp. main breaker as determined previously. This main breaker shuts off all current to the house. Purchase circuit breakers of the correct amperage. Receptacles and light circuits require 15 amps. and kitchen, laundry, and workshop circuits require 20 amps. Major appliance circuits require larger-amperage circuit breakers as specified by the appliance manufacturer. Be sure to buy 230-volt breakers for the 230-volt appliances.

Use #2 copper wire type RHW or THW for 100-amp. service. For 200-amp. service, use #3/0 copper wire. Measure the length of conduit you will need, and then buy the wires 6 feet longer. You will need one red, one black, and one white wire. Pull the wire through the conduit and the meter socket, and leave 3 feet of it extending from the entrance head for the power company to connect. The red and black wires are cut and connected to the terminals of the meter socket with red on one side and black on the other. The white wire is not cut but has a little insulation removed and is held with a center clamp.

A ground wire is required to run from the circuit breaker panel to a ground clamp. Use #6 bare uninsulated copper wire for 100-amp. service, and #4 wire for 200-amp. service. Keep the ground wire as short as possible, and use an iron clamp for galvanized pipe or a copper clamp for copper pipe. The ground clamp is usually installed on the water pipe where it comes into the house, on the street side of the water meter. If this is not possible, drive a ¾-inch galvanized pipe 8 feet into the ground, and attach the clamp to this. Check with your local electrical inspector on this as some soil conditions do not make a good ground.

The service entrance is installed either before or after the wiring of the house is done, but usually before so that the electric power will be available for drilling all the holes required.

When the service entrance is completed, one circuit breaker can be installed and a grounded receptacle temporarily connected so you can have electrical power for power tools. The power company will now connect the power.

In some cases, you will want to have electrical power before the house is framed. This is made possible by erecting a temporary pole in the yard, with a small fuse box on it. Buy an 18-foot pole or 4 x 4 and set it 4 to 6 feet in the ground. Mount the service parts as previously described. In place of the circuit breaker panel, use a small fuse box with a 30-amp. fuse. It is a good idea to padlock the box when you are away. Mount a grounded receptacle and box next to the fuse box. Install a good pipe ground connection as described above.

The next step in wiring the house is to install the electric boxes. These are 14 gauge galvanized steel with knockouts for the cables or conduit. Use the type with clamps for nonmetallic-sheathed cable, and use regular ones for conduit or armored cable. Two types of boxes are required. For switches and receptacles, use 2-¼-inch-deep rectangular boxes—they must have removable sides so two or more can be joined. The boxes are nailed to the wall studs with ½ inch protruding so they will be flush with the drywall. For ceiling lights, use 4-inch octagonal boxes, which are hung from the ceiling joists with a hanger that mounts in the center hole. Some 230-volt appliances require special receptacles which need larger square boxes. Buy the receptacles first to see if they require large boxes.

The wall outlets are centered 12 inches above the floor, and the switches are centered 44 inches above the floor. Outlets above the kitchen cabinets are also centered 44 inches above the floor. Be careful to locate the ceiling outlets in the kitchen and dining areas correctly—they are usually centered over the kitchen or dining area table.

When you are sure all the boxes have been installed, you can plan the wiring. As mentioned before, there are three wiring systems. The most common system is nonmetallic-sheathed cable, also known as Romex. Use it if it is approved by the local code. Buy type NM for indoor use, or type NMC for outdoor or damp conditions. The code now requires the use of cable with an uninsulated ground wire. Be sure you get this type, as the older type without the ground wire is still being sold. Cut the cable with a pair of large wire cutters. Remove about 8 inches of the outer cover, leaving the black and white inner insulation. If you don't have boxes with clamps in them, you will need connectors to fasten the cable to the boxes. You will also need special staples to fasten the cable to the wall studs. Ground clips are used to fasten the bare ground wire to the boxes.

If your code requires armored cable, also known as BX, it can be installed almost as easily as the nonmetallic cable. It is called type ACT, and provides better protection for the wires, but must be used in dry locations only. It is connected to the boxes with connectors made especially for this type of cable. A fiber bushing is used to protect the wires from being cut by the rough metal ends. The cable contains a bare ground wire.

Cut the cable with a hacksaw. The armor is removed by cutting through one convolution, about 8 inches from the end. Cut at about a 45° angle to the cable. When the convolution is cut through, twist the cable to break the armor and then slip it off. This becomes easy with a little practice, but be careful not to break the ground wire. Pull off the protective paper, and insert a fiber bushing between the wires and the armor. Bend back the ground wire, and slip a connector on. Tighten the screw on the connector, and then wrap the ground wire around it. Remove the lock nut and insert the connector through a hole in the box. Screw the lock nut on and tighten with a screwdriver and hammer. Use staples to hold the wire to the wall studs.

If your local electrical code requires conduit, you will have a much more difficult job. The wall studs must be drilled and notched to accept the conduit, so careful planning of all the required circuits is necessary to hold the number of conduit runs to a minimum. The conduit used is called "thin wall" or EMT and comes in 10-foot lengths. It can be easily bent with a low-cost conduit bender. The following is a chart of the conduit sizes required for various sizes and numbers of wires. If you don't feel you can figure this out, you can have an electrician install the conduit and later pull the wires through yourself.

Size of Wires	Number of Wires	Conduit Size
14	4	1/2 inch
12	3	1/2 inch
14	6	3/4 inch
12	5	3/4 inch
10	4	3/4 inch
8	3	3/4 inch

Cut the conduit with a 32-tooth hacksaw or a pipe cutter and ream the ends with a pipe reamer to get a smooth edge. Don't have more than four bends in one length. Use clamps to hold the conduit wherever necessary. The sections of conduit are joined with slip couplings and connected to the boxes with slip connectors. These are tightened with a wrench.

The conduit is installed in the walls by drilling holes in all the studs in a straight line and sliding the conduit through. To turn corners, bend the end of the conduit after it is in the wall, and notch the corner studs until the bend is flush with the surface and in line to connect to the run of conduit in the adjoining wall. The conduit on the basement ceiling or attic floor can be clamped to the joists without drilling or notching. Type TW single-conductor wire is used with conduit, and the wires are not pulled through the conduit until the walls are finished.

The first step in installing the wiring is to purchase the correct wire size. The common wire size is #14, but some codes specify #12 as minimum. In either case, the kitchen, laundry, and workshop area must have #12 wire. The special circuits for the large appliances are sized to the appliances; the amperage is usually specified on each appliance. If wattage is specified, divide by the voltage (115 or 230) to get the amperage. The following chart shows the correct wire size for various amperages.

Divide the wiring into circuits as explained in the previous chapter. Drill 5/8 or 3/4-inch holes through the studs to thread the wires. Run wires over the doors and under the windows. The wiring in an unused attic space can be stapled to the joists, but if the attic will be used for storage, you will have to drill the joists.

Crawl spaces can have the cable stapled to the bottom of the joists, but basements or garages must either have the joists drilled or the conduit clamped to the joist. Conduit gives a much neater appearance, and is easy to install in exposed areas. If the garage or basement will have finished walls and ceilings, wire them the same as the house.

Amperage	Minimum Wire Size
15	14
20	12
30	10
40	8
50	6
70	4

Wire the receptacles first using two-conductor wire. There is no set procedure to follow, as long as all the boxes are interconnected and are wired to the service entrance panel. Wire them in groups to form the circuits previously determined. To wire switches, run a two-conductor wire from the switch to the receptacle to be controlled. If a receptacle is to be controlled from two locations, three-way switches are used. To wire three-way switches, use three-conductor wire as shown in the diagrams. If two or more receptacles are to be controlled by one switch, connect the receptacles with a three-conductor wire instead of a two-conductor wire as shown in the diagrams. Split-wired receptacles are connected with three-conductor wire. All 230-volt receptacles are wired with three-conductor wire. Leave 8 inches of wire sticking out of each box, and leave 2 feet of wire where the wires go to the service entrance panel.

When you are sure all the wires are in, the electrical inspector will inspect the job, and then the walls and ceilings can be finished. Be very sure that no wires have been forgotten—check the list provided previously.

After the interior walls are finished, the switches and receptacles can be installed. At this point, the actual electrical circuits are made. The drawings show most common connections. Remember to keep all the black wires together and all the white wires together. When wiring a switch, both wires are considered as an extension of the black wire in the receptacle controlled. To avoid future confusion, the white wires to the switches should be wrapped with black tape so none of the white shows on either end. Red wires used in three-conductor cables are considered the equivalent of black wires, so they do not need to be covered. The white wire will be grounded at the service entrance panel, and is called the "common" wire. The black and red wires are called the "hot" wires.

Probably the biggest problem encountered in wiring a house is the realization, after the walls are covered, that the wires can no longer be seen. You can not tell any more where all the wire ends are connected. This can be eased by marking the ends of each wire with colored tape or tape with numbers on it before the walls are covered. It is still possible to lose track of where they go, but this is actually easy to remedy. Take the two wires coming from one cable and connect the wire ends with a solderless connector. Next, find what you believe to be the other end of the cable. Buy or make a test unit with a 1½-volt flashlight battery and bulb. Touch the wire ends of the test unit to the wire ends of the cable—if the bulb lights, you have the correct cable.

Wires are never spliced or connected anywhere except in a box. To connect groups of wires, clean about 1 inch of insulation away and twist the wires together with a pair of pliers. It is best to solder the wires and it is also very easy to do. Use a small propane or butane torch and rosen core solder (do not use acid core solder). Heat the wires with the torch and then touch the solder to them; wrap the connection with plastic electrical tape. Another way to connect the twisted wires is to screw on a solderless wire nut, available in various sizes for different sizes and numbers of wires. Be sure to twist the wire clockwise. This may sound simple, but soldering and taping is actually easier to do correctly and provides a much more dependable and safer connection.

Connect the receptacles first. There are two types of duplex receptacles. One has two screws on each side for connecting the wires coming to and leaving the box. When using this type of receptacle, make sure the wire is wrapped *clockwise* around the screw so it will be coiled tighter when the screw is tightened. Another type of receptacle has holes to insert the wires into; no screws are used. These are fine and easier to use, but may cost slightly more. Special receptacles, with hinged covers and gaskets, are required for outdoor outlets.

Note that the screws are silver on one side of a receptacle and gold on the other side; always connect the white wires to the silver screws and the black wires to the gold screws. Grounded

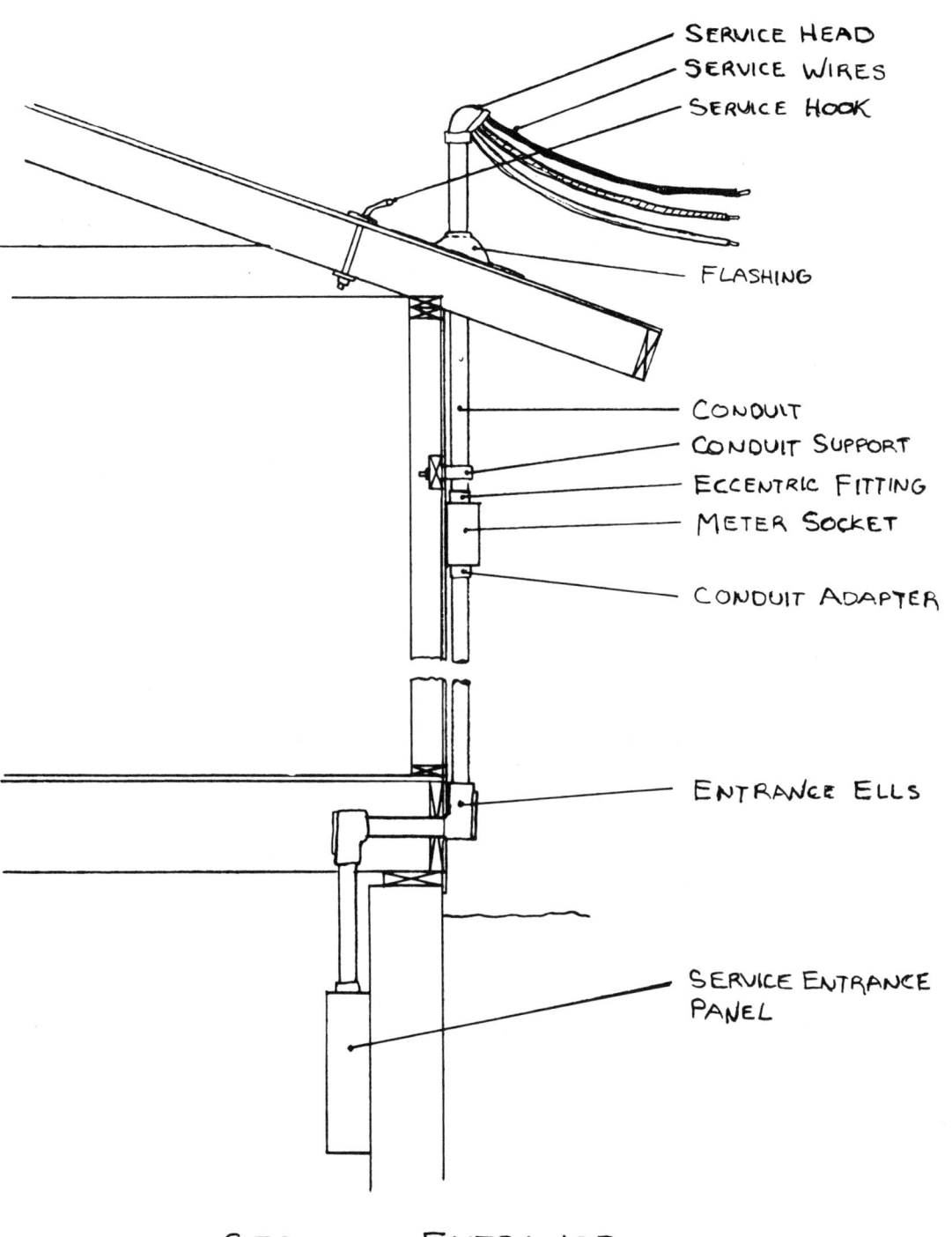

SERVICE ENTRANCE

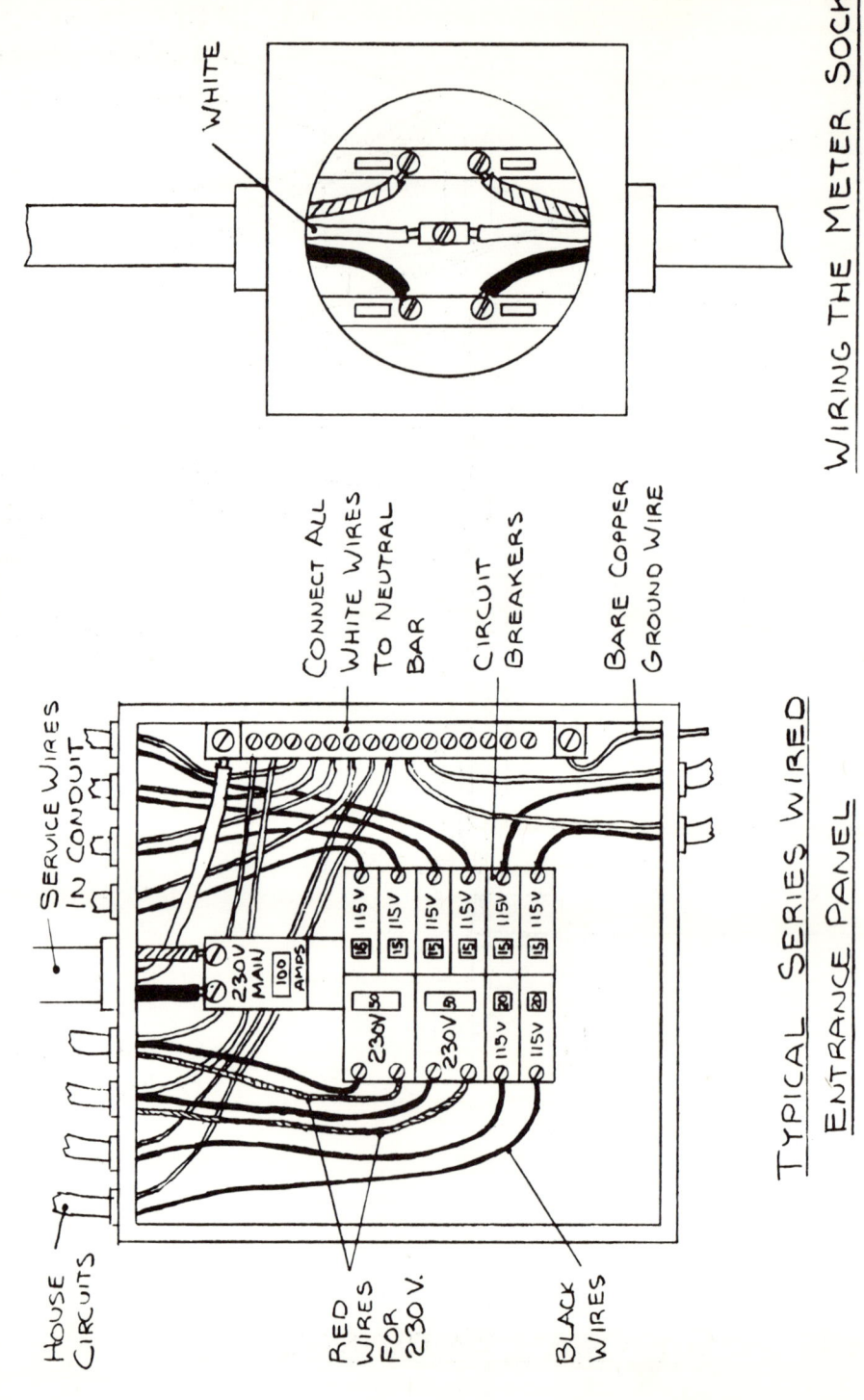

ARMORED CABLE   CONDUIT

CONNECTORS

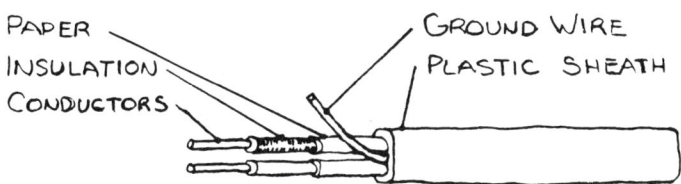

NON-METALLIC SHEATHED CABLE (ROMEX)

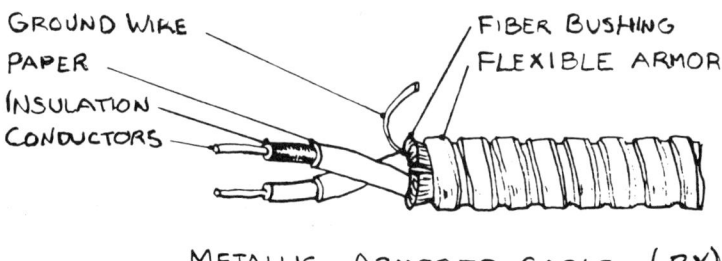

METALLIC ARMORED CABLE (BX)

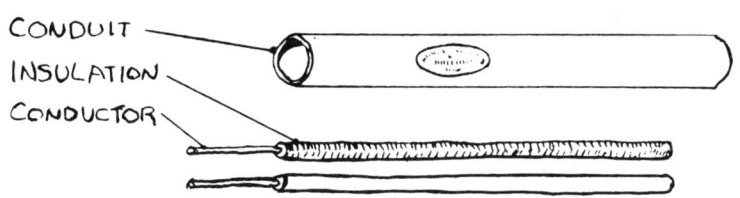

CONDUIT AND INSULATED WIRE

THE THREE WIRE SYSTEMS

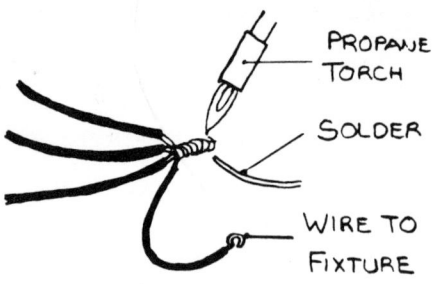

METHOD OF JOINING WIRES

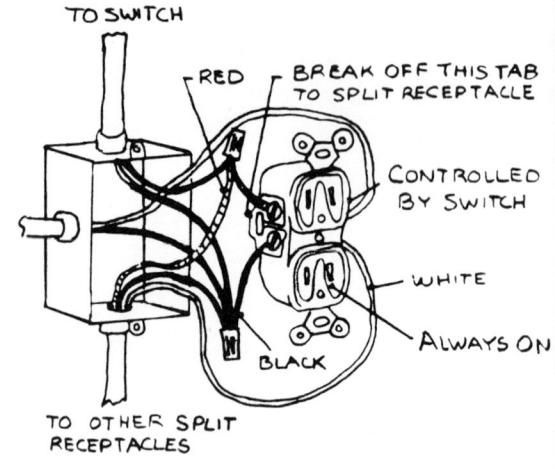

SPLIT-WIRED RECEPTACLE

SOLDERLESS WIRE NUT

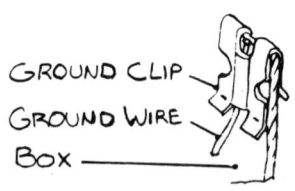

GROUNDING CLIP

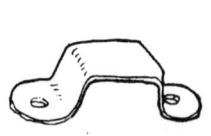

CABLE CLAMP

STAPLE

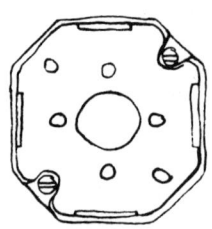

OCTAGONAL BOX
FOR LIGHT FIXTURES

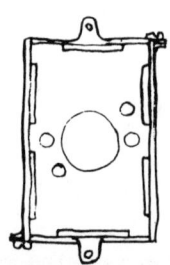

RECTANGULAR BOX
FOR RECEPTACLES & SWITCHES

Conventional Switch Wiring

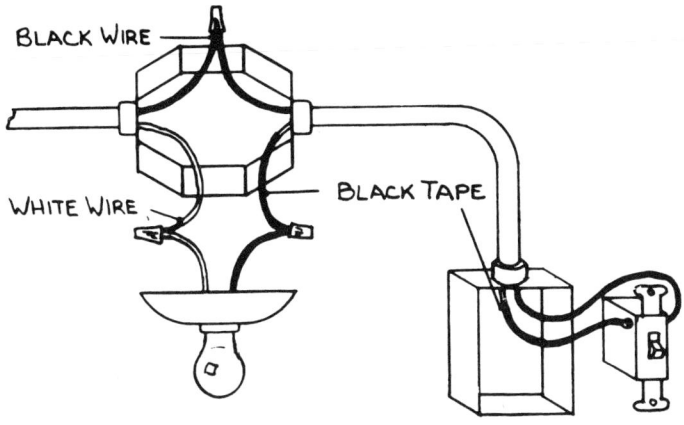

Switch Beyond Receptacle

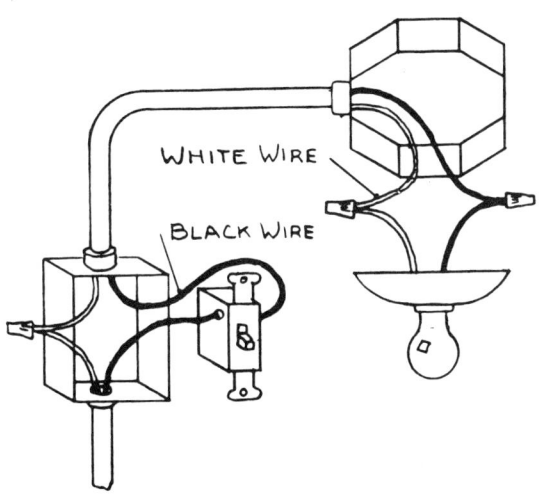

Receptacle Beyond Switch

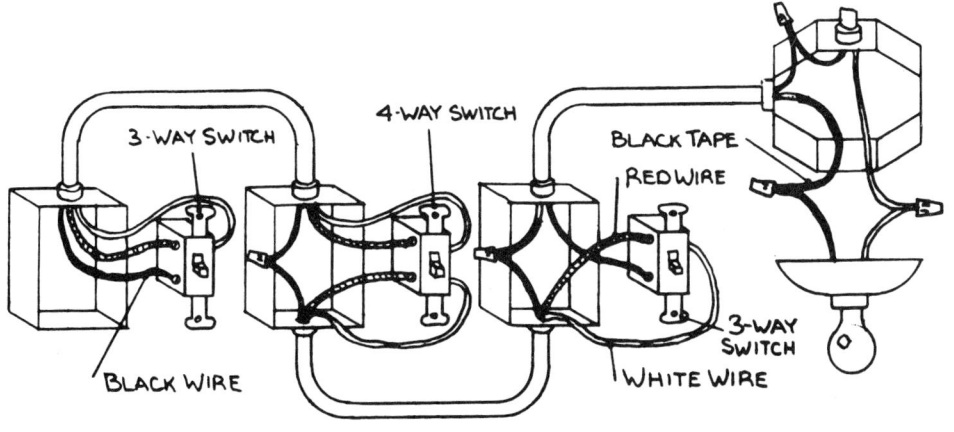

Four-Way Switch Wiring

THREE-WAY SWITCH WIRING

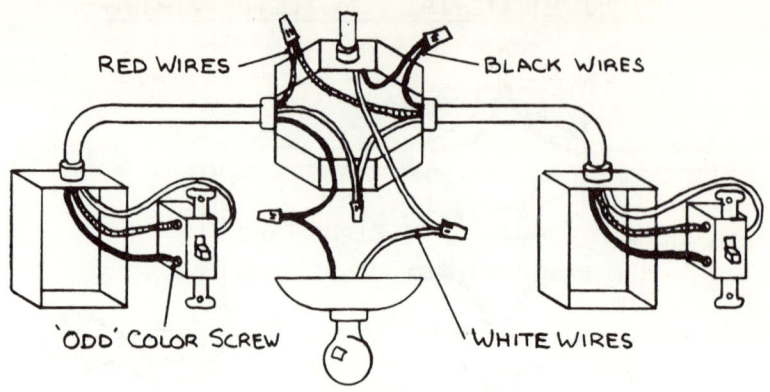

RECEPTACLE BETWEEN SWITCHES

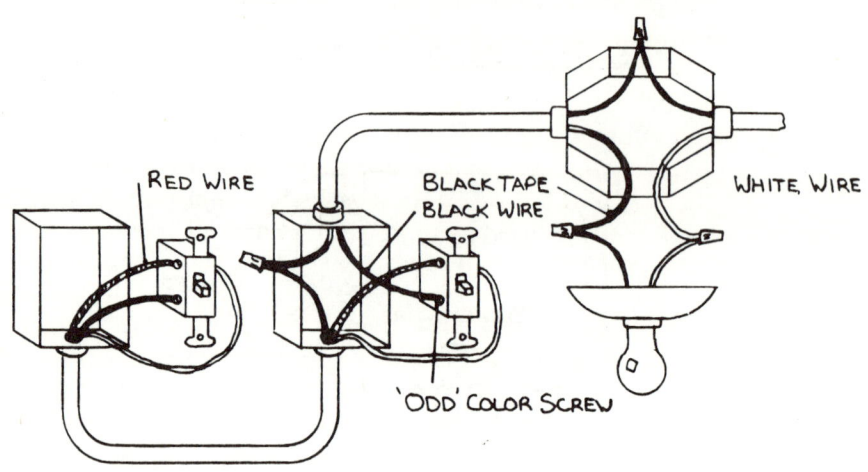

SWITCHES BEYOND RECEPTACLE

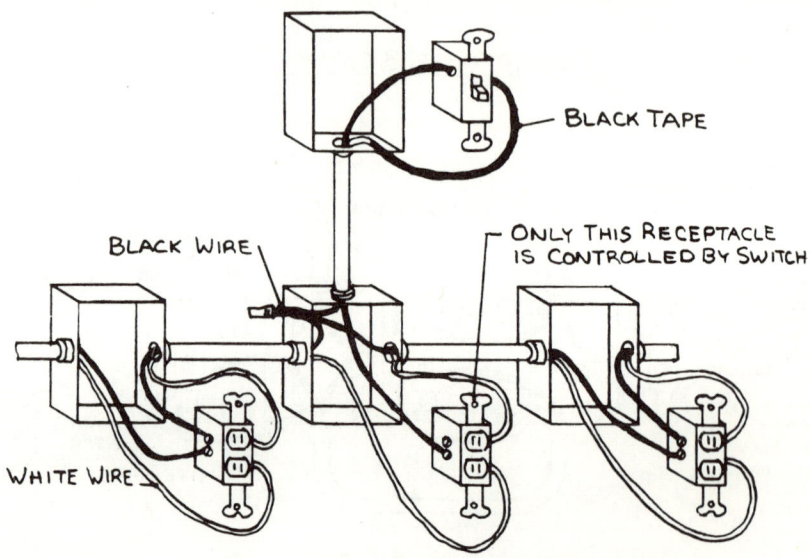

WALL RECEPTACLE WIRING

receptacles should always be used in the basement and garage and on all the 20-amp. circuits such as the kitchen and laundry. These receptacles have a third opening for the grounding prong used on grounded-appliance cords. They also have an additional green screw which is grounded to the metal box with a wire and screw or grounding clip.

Light fixtures have either two screws, like the receptacles, or two wires. The wires are either one black and one white, or one with silver wire and one with gold wire. Connect black to gold and white to silver using solderless connectors. To split a receptacle, break off the removable tab on the side with the gold screws. This separates the upper and lower receptacles. Connect the white wires to the silver screws, and the black wires to the lower gold screws. Connect the red wires to the upper gold screws. The wires from the switch controlling the upper receptacles are connected to the gold screw. In all cases, cover the white wire on the switch with black tape so it will not be confused with a "common" wire.

Wiring of 230-volt appliances requires special receptacles, which come in various amperages. Use the same amperage receptacle as used for the circuit breaker. When all the switches, receptacles and light fixtures have been wired, wire the service entrance panel. If the electricity has been connected by the power company, you must have them temporarily disconnect the meter again.

Install the circuit breakers in the box. Connect all the black wires from the circuits to the correct circuit breakers. Connect all the red wires from the 230-volt circuits to the extra screw on the 230-volt circuit breakers. Connect all the white wires to the silver neutral bar. If you are using nonmetallic-sheathed cable, run the bare ground wires to the neutral bar, also.

Have the power company reconnect the power. If one of the circuits is wired incorrectly, one of the circuit breakers will trip. If it does, shut off the main breaker, and recheck the wiring on the defective circuit.

36. Insulation

After the house is closed in and the electrical work is roughed in, the insulation can be installed. Use paper- or foil-backed fiberglass batts. All walls and the ceiling should be completely insulated, and if you have a garage, you may want it to be insulated too. Insulating the ceiling will keep the garage cooler in the summer; insulating the walls will keep it warmer in the winter. If you plan to have a heater in the garage, be sure both the ceiling and walls are insulated.

Ceiling insulation should be at least 3-½ inches thick. If you plan to have electric heat or air conditioning, it should be 6 inches thick. Insulation is low in cost so use enough, especially in very hot or cold climates. Remember: If the house has no attic, or a shallow attic, it will be hard to add more later. Ceiling insulation is easier to install if the batts are cut into 6-foot lengths. Use a staple gun to fasten the paper backing to the underside of the ceiling joists—the paper or foil side is always toward the room. The insulation must extend down over the top plate of the exterior walls, but be sure you don't block off the air flow between the joists. Butt the pieces tightly together, and fit tightly around electrical boxes.

Exterior wall insulation is usually 3-½ inches thick. Sometimes 1-½-inch thickness is used, but only in very mild climates without electric heat or air conditioning. Here again, be sure to use enough, because once the walls are finished, you cannot add more. Just as with the ceiling, the wall insulation is stapled to the studs with the paper or foil side toward the room. Cut the batts to the correct length first; fit tightly around all electrical boxes and stuff some scraps behind the boxes. Fill in all cracks between the framing and the window and door frames, no matter how small they are.

The wood framing around the windows and doors should be

covered with paper or foil to form a vapor barrier. When the insulating is done, the entire wall and ceiling surface of the house should be covered with vapor barrier.

In very cold or hot climates, extra insulation may be required in brick veneer houses. Brick is not a good insulating material, despite what many people believe. It would require 5 feet of brick to equal the insulating value of 3-½ inches of fiberglass. Insulation board exterior sheathing is commonly used under brick veneer—it is installed in place of the plywood sheathing. Leave a full sheet of plywood on each end of each exterior wall to stiffen the walls. The rest of the exterior can be insulation board, which will help make up for the heat loss caused by the brick veneer. It will also help keep the house cool in the summer, insulating against the heat built up in the bricks by the sun.

37. Wall and Ceiling Finish

A variety of wall and ceiling finishes are available. Plaster, which was once the only acceptable material, is rarely used any more, and since plaster requires skilled labor, it will not be discussed here. The most common material used today is gypsum drywall. It is fairly easy to put up, although the seams usually must be finished by an expert. One big advantage of gypsum drywall is that it is fireproof. Prefinished plywood is an excellent wall material, and can be installed over the studs or over drywall. Acoustical ceiling tiles make an excellent ceiling and reduce the noise level in the house.

Gypsum drywall comes in two thicknesses: ⅜-inch is commonly used for ceilings and ½-inch for walls. However, either thickness is approved for walls or ceilings. Nail with special gypsum drywall nails—4d for ⅜-inch thickness and 5d for ½-inch thickness. Drywall is always installed with the finished edges at right angles to the joists or studs, and it is available in 8- 10- and 12-foot lengths, to minimize seams. Nail every 8 inches on all frame members and dimple all nails. Gypsum drywall is very heavy material so you will need help in installing it, especially on the ceilings. Make two tee-shaped supports ¼ inch taller than the ceiling to hold the sheets while nailing. A special hammer is available to do a better dimpling job. Special waterproof drywall must be used around bathtubs—never use regular gypsum drywall or it will disintegrate from the moisture. If you plan to have the seams taped and finished by a professional, consider having the drywall installed also since finishers prefer doing the whole job, are more expert at it, and get better results.

If you plan to tape the joints yourself, follow the instructions which come with the tape and joint compound. Use metal edges on all outside corners. Practice in an area which will not show, such as a closet. Have good lighting available so you can see any defects. With lots of practice, you can do a presentable job; but if you want perfection, you may decide to hire a professional.

Acoustical ceiling tiles, if you wish to use them, can be cemented to the gypsum board. Get the correct cement and follow the instructions. Many types of acoustical tile are available, some of which have a vinyl surface to make cleaning easier. Acoustical tile installed over gypsum board makes a good-looking and sound-absorbing ceiling, and eliminates the need to do a good job taping the drywall joints.

If you plan to panel the walls, choose a good grade of paneling ¼-inch thick. Paneling also sometimes comes with a vinyl coating for a lasting finish. Choose a wood and color that will look nice with your furniture or other walls. Paneling can be applied over drywall or directly to the wall studs with adhesive and nails. Place panels against the walls, starting in the most noticeable corner, and match the wood grain for the best effect. Trim the edge of the first sheet as required to fit in the corner—the edge of the panel must come exactly ¾ inch from the edge of the stud. Apply adhesive to the studs and the plates as recommended by the manufacturer. Level the panel and nail along the top first. Use 3d finishing nails and space the nails 8 inches apart on the edges and 16 inches apart in the center. Set the nails about 1/16 deep and fill the holes with a matching color putty stick. Measure carefully for windows, doors, and electrical outlets.

Use matching trim moldings, and use corner beads over outside corners. Cove and inside-corner molding are used in inside corners; crown or cove molding is used around the ceiling. Casing is used around the windows and doors, and base is used along the floor. (See *Interior Trim*.) Paneling installed over gypsum board is and easy way for an amateur to do a professional-looking job.

Ceramic tile is the most common bathroom wall finish for areas which are exposed to moisture. Ceramic tile generally is either 4 x 4 inch squares or 1 x 1 inch mosaic type. There are many others, such as 2 x 2 inch and ½-inch Italian tiles. The mosaic tiles, which come in 12 x 12 inch sheets, are the easiest to install. Mark the outline of the area to be tiled on the wall. Mastic which can be cleaned up with water is available and makes the job much less messy.

Spread the wall tile mastic on the walls with a special notched trowel and put the tiles in place, sliding and twisting them until they are in the proper position. Then press them firmly in place.

Lay them in a way that makes it unnecessary to cut the tiles to less than one-half their width. Use ceramic tile pliers to cut the mosaic tiles, or if you are using larger tiles, borrow a tile cutter from your dealer. This is a tool which holds the tile, scribes it, and has a lever to break it evenly. After the adhesive is dry, mix the white grouting and spread it with a trowel. Force it into the cracks with the handle of an old tooth brush; let it dry for a few minutes, and then wipe the surface clean. Strike the joints again and wipe clean again. Be sure the tile is perfectly clean before the grouting is dry. In about a week, after the grouting is completely dry, coat the tile with a silicon waterproofing available at your tile dealer.

Garage walls may be covered with ½-inch gypsum drywall. If the garage is not going to be finished, the house wall and ceiling must be covered with ½-inch gypsum drywall for fire protection. If no ceiling joists are used in the garage, the house wall must be covered with ½-inch gypsum drywall up to the roof sheathing.

38. Cabinets

Kitchen and vanity cabinets are available ready made and finished, and in a wide range of prices and styles. They come in modular sizes and can be combined to fit any kitchen plan—most house plans show arrangements for the kitchen. Make sure you can use standard-size cabinets to keep the cost down. The counter tops can be purchased ready made to fit, with finished ends and a hole cut out for the sink. The straight sections used on many plans will keep the cost to a minimum. Most homes have bathroom vanities, which can be purchased to match the kitchen cabinets if desired. Notice that the vanities are lower and not as deep as kitchen cabinets. Vanity tops can be purchased with built-in wash bowls. (See *Plumbing Fixtures*.)

You can also build your own cabinets. The cabinet plans show typical modular-sized cabinets which can be built and fastened together; or the whole wall of cabinets can be built as a unit.

The materials are interior plywood and clear pine boards with prefinished plywood fronts. Each cabinet has the front surface extended ¼ inch on each side so that the cabinets will fit tightly together. This also provides room for a ¼-inch prefinished plywood panel for appearance' sake if the end of the cabinet is exposed. The top can be purchased or made from scratch.

First, build the base cabinet. The base is made from two 1 x 4's cut 19 inches long and two cut 1-½ inches less than the cabinet width. Nail these together with the joints glued with good white glue, such as Elmer's Glue-All. Cut out the ½-inch plywood end panels 23 x 34-½ inches with a 3-¼ x 3-½ inch toe space, and nail and glue to the base. Cut two 1 x 2's 19 inches long and two 1-½ inches less than the cabinet width and nail and glue to the tops of the plywood. Next glue and nail the ¼-inch plywood or ¼-inch hardboard back in place. It measures 34-½ inches high by ½ inch less than the cabinet width. The shelf is a 1 x 12 resting on a 1 x 2 frame, and the floor is ½-inch plywood.

The front frame is made from two 1 x 2's 31 inches long and a 1 x 2 and 1 x 3 each 3 inches narrower than the cabinet. Nail and glue these together, being careful to see that they are square. Another 1 x 2 is used below the drawer area—leave 4 inches for the drawer. If the cabinet is over 24 inches wide, it will have two drawers and two doors with a 1 x 4 used as a divider in the center. Fasten this front frame to the cabinet and glue. Use finishing nails on this since it will show, and set the nails 1/16 inch below the surface.

The drawers are made of 1 x 4's with a ¼-inch plywood bottom. The sides are cut 20 inches long, and the front and back are ¼ inch narrower than the opening. The doors are made from ⅜-inch plywood, ¼ inch smaller than the opening both ways. The fronts of the doors and drawers are made of prefinished ¼-inch plywood, although they may also be made of pine boards or regular plywood and painted. With a little imagination, very beautiful fronts can be made. Omit the drawers below the sink or lavatories and cover the openings with a matching panel. If the end of a cabinet is exposed, cover it with matching wood.

The cabinet top can be made of ¾-inch plywood with a 1 x 2 frame around the underside. Leave a 1-inch overhang on the front and each exposed end. To cover with formica, glue the top in place first, with a ⅛-inch overhang on the front and finished ends. Then butt the 1-½-inch-wide edge strips up under the top. A special routing tool is used to trim the top flush with the edges. The back splash, if one is desired, is done the same way. It is made by covering with formica a 1 x 4 board on one side and one edge; screw it to the top with long screws up from the bottom. Your formica dealer will have the proper glue and routing tool for the covering job. Before making your own top, however, check the prices on ready-made tops. If you use a standard color, the price is sometimes lower than the cost of making your own.

The wall cabinets are made in the same way as the base cabinets. The vanity cabinets are also about the same as base cabinets, except that they are 32 inches tall (including the top) and 21-½ inches deep. If wall cabinets are used in the laundry area, they are the same as the kitchen cabinets.

The drawers are held up by drawer guide sets, which can be purchased, and which have either nylon guides or rollers for smooth action. The doors are hung with special cabinet hinges, available at any hardware or lumber dealer. A wide variety of knobs and handles are available for the doors and drawers.

The finished cabinet can be either stained and varnished, or painted. If the doors and drawers are prefinished plywood, the front frame can be stained to match, or painted black. The top space is usually painted black or covered with a base molding.

Attach the wall cabinets first by screwing through the upper shelf and bottom frames into the wall studs. If the cabinets are to be ceiling hung, they are screwed into the soffit, or dropped ceiling. Use screws large enough to hold a fully loaded cabinet.

The base cabinets are also usually attached to the wall. If the cabinets are free standing, they must be fastened to the floor. This is easier if the bottom plywood is left out until the cabinets are screwed to the floor. Some 1 x 2 strips can be glued to the frame to nail through into the floor.

If you plan to have a dishwasher, it can be built in by eliminating one 24-inch base cabinet next to the kitchen sink and extending the counter top across the opening.

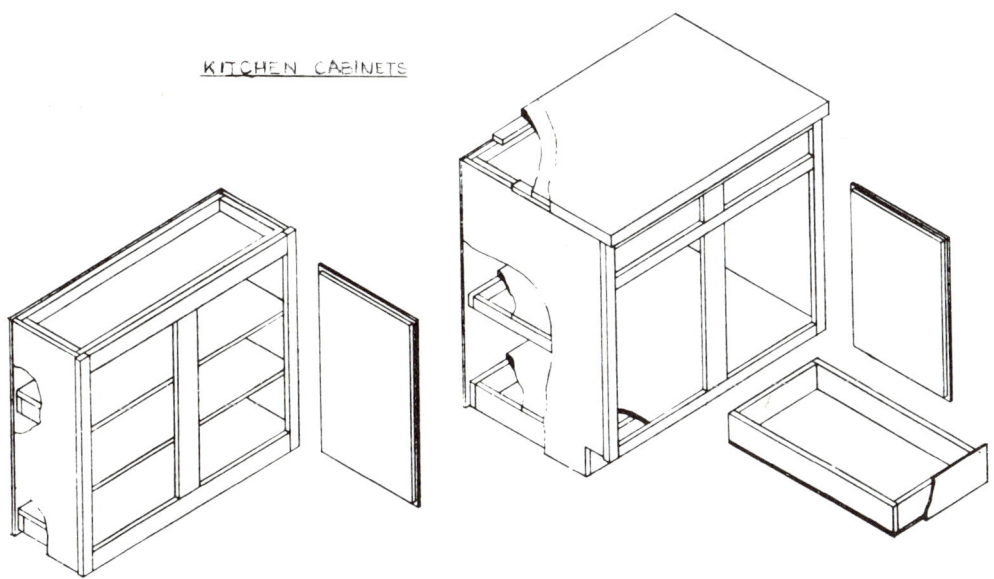

KITCHEN CABINETS

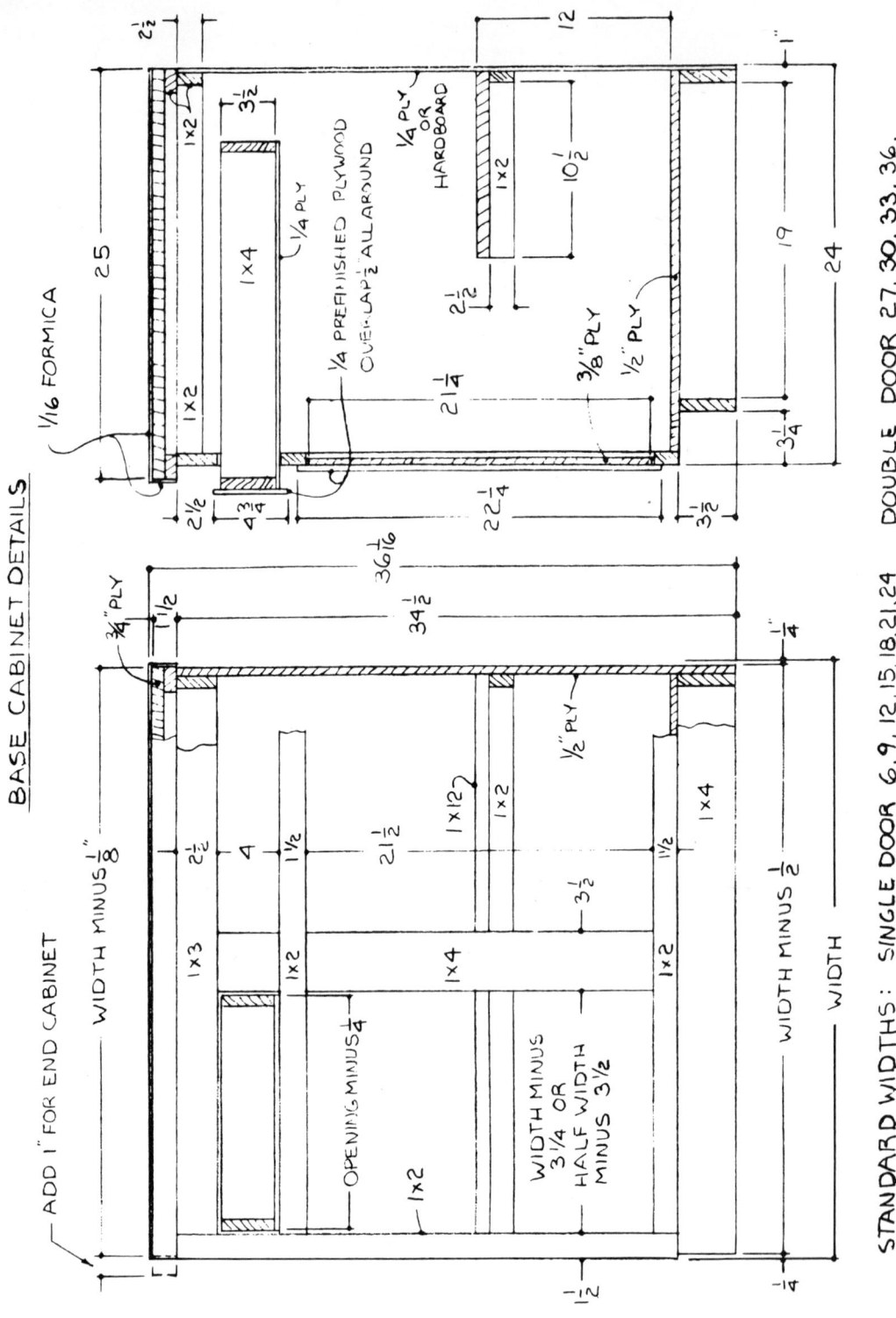

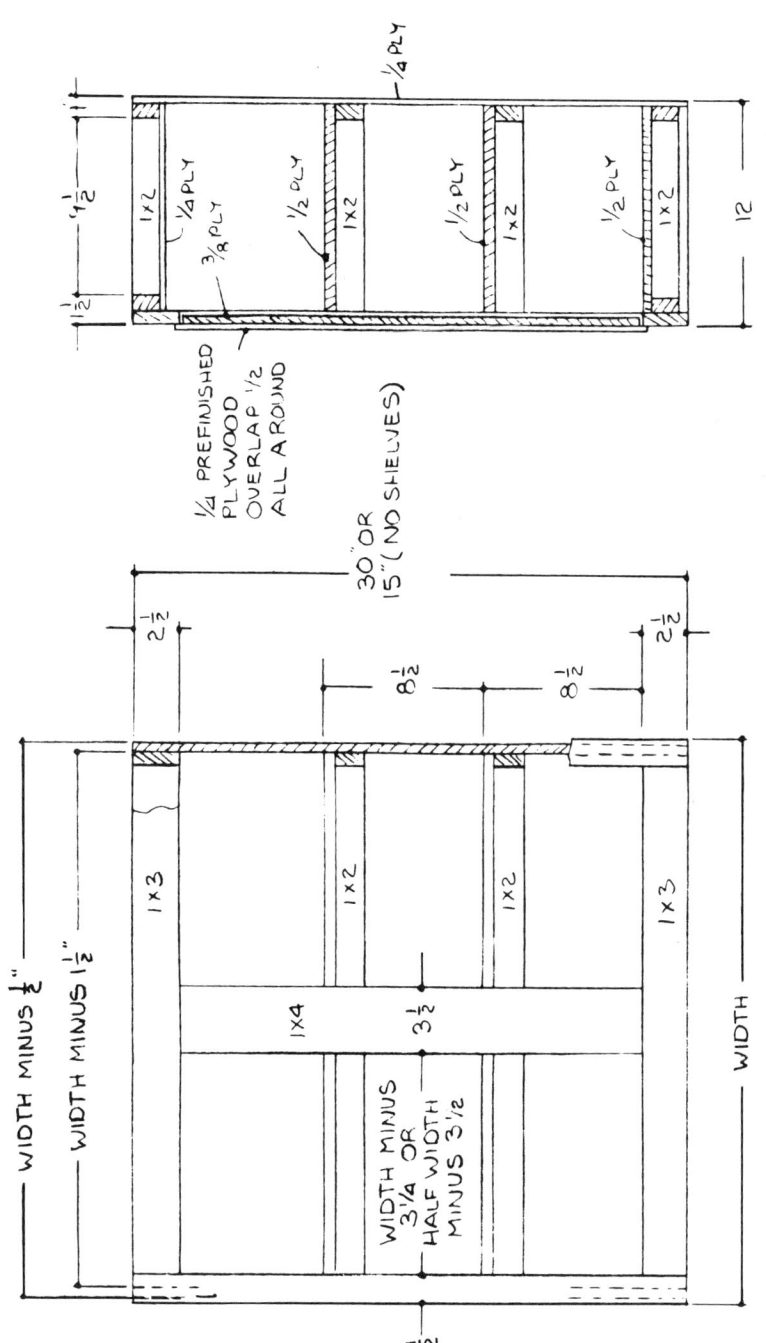

BATH & KITCHEN ELEVATIONS

KITCHEN

D	E	F	G	H
8'	6	30	OMIT	24
9'	12	36	OMIT	30
10'	18	24	18	36
11'	24	24	24	36

LAUNDRY

KITCHEN

A	B	C
8'	15	15
9'	24	18
10'	36	18
11'	36	15

SMALL BATH

LARGE BATH

39. Interior Trim

After all interior walls are completed, the trim can be installed. If the walls are prefinished plywood, you may want to stain the door and window jambs first, and then apply prefinished trim. If the walls are to be painted, and the trim stained, you may want to paint the walls first, then fit the trim and remove and stain it before nailing it permanently in place. If both the walls and the trim are to be painted, install the trim after the walls are sanded.

Modern trim is smaller and plainer than that used years ago; only four basic types are used. Doorstop trim is used when hanging the interior doors. It measures ½ inch by 1-¼ inches, and should be nailed to the door jamb with 4d nails 12 inches on centers. Casing is used as trim around all doors and windows. It measures ½ inch by 1-¾ inches and is used with the rounded or thin side toward the door or window. Nail with two 4d finishing nails every 16 inches—one goes into the door or window jamb, and the other into the framing. Base is used around the base of the walls, and is installed after the door casings. Base measures ½ inch by 2-½ inches and is nailed to the bottom plate and the studs with two 6d finishing nails every 16 inches. Install the base after the underlayment or hardwood flooring is laid, but before carpeting or ceramic tile. For use with linoleum or vinyl floors, the base can be loosely fitted, removed until the flooring is installed, and then renailed in place. This may eliminate the need for shoe molding, which is used to cover any gap between the base and the finish flooring. It is never used with carpeting or ceramic tile. Shoe molding measures ½ inch by ¾ inch and is nailed into the base with 4d nails spaced 16 inches apart.

Corners of trim can be cut in a wooden miter box, or in a metal miter box with saw guides. If the trim is to be painted and the corners do not come out perfectly, they can be filled with spackling.

Some windows come with a protruding stool. If you have this type, the casing comes down flush with the top of the stool, and a piece of casing is installed thick side up, under the stool. If the window has a flush stool, the casing goes around all four sides like a picture frame. All casing is installed ¼ inch back from the edge of the jamb.

When doors are hung (see *Doors*), they will be left long until the finish floors are installed. They are then cut off to provide clearance over the finish floor. Leave ⅜ inch over carpeting and ¾ inch over other floors to clear throw rugs. The edges of the door should also be planed on a slight angle to provide ⅛ inch clearance on the top and opening side. The clearance on the hinged side is controlled by the hinge design. Install aluminum thresholds under exterior doors; these usually come in 3-foot lengths and can be cut to fit. They also have replaceable vinyl inserts. Cut the exterior door to compress this vinyl slightly to give a close seal. The top and opening edge of the exterior door should be planed to contact the weather stripping without too much friction.

40. Finish Floors

Finish flooring materials are a matter of personal choice. However, the trend is definitely to carpeting for all rooms. Other popular materials are vinyl or slate in entry halls, ceramic tile in baths, and vinyl or linoleum in kitchens. Hardwood is still used in some areas, especially in authentically styled Colonial or Early American houses, but it is somewhat impractical by today's low-maintenance standards.

Underlayment is required for all but hardwood floors. The underlayment is ⅜-inch interior plywood of a grade which provides a smooth top surface. Nail to the subfloor with 3d annular or screw thread nails, and set the nails below the surface to allow for shrinkage. Nail 8 inches on centers all around the edge and over the entire surface. Be sure the subfloor is clean and smooth before applying underlayment—use a vacuum cleaner if possible.

Carpeting can be installed directly over the underlayment. Regular wall-to-wall carpeting laid on a resilient pad is installed by using a carpet stretcher. This can be done by any carpet dealer. Indoor-outdoor carpeting of the loop pile variety with a foam backing or carpet tiles can be laid by anyone. Your carpet dealer can give you all the necessary information. This is a very good flooring material—it can be used in every room, including baths and kitchens, it wears well, and has the lowest upkeep. Carpeting usually is considered part of the house and therefore is covered by most mortgage loans.

Linoleum and sheet vinyl provide a smooth, seamless floor. The vinyl requires very little upkeep and usually does not need waxing. It is generally installed over a felt underlayment. This is a tricky job because of the large size of the sheets, usually 6 or 12 feet wide. If you wish to try this, you can get all the information you need from your local dealer. You can have it installed by the dealer, of course.

Vinyl or vinyl asbestos tile in 12-inch squares is much easier to install. Some types are thin and must be laid over a felt underlayment, although others are thick enough to lay directly on the plywood underlayment. Use contact type adhesive, if possible, since it is far less messy than other types and can be applied with a paint roller. After the adhesive is applied, it is left to dry to a tacky state. The tiles are then simply laid in place, and when the positioning is correct, pressed firmly in place to seal. When laying tile, always start in the center of the room and work out. Check first to be sure the last tile against each wall will be a practical width. Make a chalk line down the center of the room the long way and apply the adhesive up to the line. Do one half of the room first, then the other half. The tiles around the edge are easily cut with scissors or sheet metal shears. Most tile dealers will give you an illustrated booklet on how to lay the tile.

Ceramic tile or slate flooring is easily installed using the modern adhesives and vinyl-base groutings. The ceramic tile comes in 12-inch-square sheets. Spread ceramic floor tile mastic on the floor with a special notched trowel and put the tiles in place by sliding and twisting them until they are in proper position. Then press them firmly in place. The tile can be easily cut with tile cutting pliers or a tile cutter which you can borrow from the tile dealer. After the adhesive is dry, mix the vinyl-base grouting and spread it with a trowel. Force it into the cracks with the handle of an old tooth brush. Let it dry for a few minutes and then wipe the surface clean. Strike the joints again and wipe clean again. Be sure the tile is perfectly clean before the grouting is dry. Ceramic tile is the best flooring for bathrooms, as it is completely waterproof. Carpeting can be used over it if a warm floor is desired, but never use carpeting over an unfinished plywood floor in the bathrooms because the water it absorbs will rot the plywood.

Hardwood flooring is no longer difficult to lay. It is available in Oak, Maple, Beech, and Birch, but Oak is most common. Widths vary from 1½ to 2¼ inches. It comes in bundles of short pieces and is tongue-and-grooved and end-matched. Be sure the kind you buy has the nail holes predrilled. Clean the floor and put down a layer of 15-pound asphalt-impregnated felt. The first strip is nailed against the outside wall, with the tongue toward the room, using special 8d flooring nails. Face nail this strip along the outside edge, being sure the base molding will cover the nails. Drive each piece up tightly against the previous one before nailing. Use the longer pieces for the centers of the larger rooms. Rent an electric floor sander from a paint store to sand the floor before finishing, although prefinished flooring is also available to eliminate the sanding and finishing operations.

Brick or stone floors can be laid for the front entry, but this must be planned in advance. The area to be brick must be lower than the surrounding floor, so the floor joists must be cut down to allow the subfloor to be lowered the thickness of the stone or brick. This will require extra floor joists to support the joint between the adjoining walls and the lowered floor. Set the brick or stone in place using ceramic tile adhesive, but leave ⅜-inch spaces for the mortar. After the adhesive is dry, mix the mortar and force it into the cracks. Smooth the joints very carefully so the floor will be smooth and easy to clean.

41. Painting and Wallpapering

Painting is a very easy job if done with the modern acrylic and latex base water paints. The trend, however, is toward low-maintenance houses which eliminate as much painting as possible.

Exterior painted surfaces should be primed with a good quality alkyd primer. The finish coats should be acrylic exterior-grade paint, which is moisture-resistant and fast-drying. It is water-thinned so the brushes and rollers can be easily washed out. Many new homes are being covered with stained siding, which does not require as much upkeep as painted siding. If you are going to have this kind, consider having it pressure- or dip-stained before it is put up—most lumber dealers can arrange this.

Interior house paints are usually latex based, and custom-tinted to almost any color you can imagine. These water-thinned paints are fast-dryers, and have a pleasant smell, too. Latex enamel which has a satin gloss finish is best for use in kitchens, bathrooms, and on all wood trim. Alkyd base primers are used for most wood surfaces, but new latex primers are also available. Concrete or wood floors can be painted with oil base porch and deck enamel—latex paint is also available for floors, but the dull surface gets dirty very fast. Interior trim can also be stained and varnished. If you want a professional job, you may want to hire a painter to do this.

Wallpaper is making a comeback, and many new homes have a few wallpapered walls in combination with painted or paneled walls. Most wallpaper now comes pretrimmed and is butted together to make invisible joints. Prepasted paper which is simply soaked in water to prepare for hanging, is available.

To hang wallpaper, start in a corner. Using a chalk line and plumb bob, make a line on the wall 1 inch from the corner for a starting point. The line is placed to allow the first strip to extend around the corner. Unroll the paper on a table for measuring and cutting. Add 4 inches to the height of the wall for trimming; if the paper has a pattern, add the height of the vertical pattern to allow for matching.

Paste the paper on the table and fold both ends toward the middle for carrying, being careful not to crease the paper. Unfold one end and line up the edge with the chalk line. Starting at the top, smooth out the paper with a wallpaper brush; work from the center to the edges to remove bubbles. Crease at the ceiling and baseboard and cut with a razor blade knife. Hang the next strip by butting and matching the pattern—be careful the strips do not overlap. Roll the seams with a seam roller. In corners, measure the width required, then cut the paper 1 inch wider and wrap around the corner. Start the next strip by using a chalk line to get the strip straight, and overlap the 1-inch piece from the other wall. Around door or window casings, cut the strips 1 inch wider than the space and crease into the corner. Trim the paper by running a razor blade knife in the corner. Wipe away excess paste as you go with a dry rag, or a damp sponge on washable paper. Cut the last piece to overlap the 1-inch piece left in the corner by the first one.

42. Appliances

Selection of built-in appliances is an important part of the design of a house. The following appliances should be considered before building.

The kitchen range and oven are sometimes built-in, but not always. A separate oven cabinet is expensive and uses up 24 inches of valuable counter space. The range cabinet is also an additional expense. The modern two-oven free-standing ranges with one oven at eye level cost less than the built-in variety and take up less space. They are designed to fit snugly into a 30-inch space between cabinets. Electric stoves require a separate 230-volt circuit breaker and electrical circuit; gas ranges require gas connection and a 115-volt receptacle. Also, consider the new ultrasonic ranges which cook with microwaves.

The range should have a range hood with a power fan to exhaust steam and smoke to the outdoors. Some ranges have a built-in hood.

The space left in the kitchen for the refrigerator should be 36 inches wide and 69 inches high, which is large enough for the 20- to 22-cubic-foot models with side-by-side refrigerator and freezer. The refrigerator requires a 115-volt receptacle. You also may want to have a separate freezer in the basement or garage.

The kitchen sink can be of any type, almost always with a double basin. A garbage disposal can be easily added by installing an electrical receptacle inside the cabinet.

A dishwasher can be added by providing connections for water and sewer, and an electrical receptacle. The dishwasher can be installed later by eliminating a 24-inch section of the cabinets. Some units have their own cabinet top and others slip under the kitchen cabinet top.

Automatic washing machines must have connections provided

for hot and cold water, sewer, and electrical receptacle. Dryers are either electric or gas. The electric models are usually 230-volt and require their own circuit breaker and electrical circuit; the gas models require 115-volt electrical receptacle. Houses without basements should include a closet space of 30 by 60 inches for an upstairs laundry. If you are building a vacation house or have a small family, you may want a combination washer and dryer to save space.

If you have a window air conditioner, provide a receptacle under the window you will use. If the air conditioner requires 230-volts, provide a separate circuit and circuit breaker. If you don't already have an air conditioner, consider central air conditioning. You can install it yourself for under $500 if you have forced air heat, and it is much quieter and more efficient than window units. Intercoms, some with built-in radios, are also available, and the wires should be put in the walls before they are finished.

Incinerators for burning garbage and papers usually operate on gas. They can be installed in the garage by connecting the gas and installing a prefabricated chimney.

43. Concrete Flatwork

Concrete walks, drives, the patio, and garage floor can be easily laid. Basement floors and garage floors should be done by an expert cement finisher if you want a good finish, but if you have experience or will accept an average job, you can do it yourself. The floor slab for the house should be perfectly smooth and level, so unless you are an expert, have a cement finisher do it.

Walks and drives are laid by measuring the area and marking it with stakes. Make sure the corners are square by measuring the diagonals and seeing that they are equal. Build a form with 2 x 4 boards and 2 x 2 stakes; the stakes go on the outside and hold the 2 x 4 forms straight. Level the area by filling it with gravel even with the bottom of the forms. If the form butts up to a fixed object such as the garage floor or a foundation, install an expansion joint, which is available from lumber dealers. Stand the joint up against the fixed object. Cover the entire area with 6 x 6 inch reinforcing wire.

The next step is very important and often overlooked. Where walks, drives, or patios come up to the outside wall of the house, provision must be made to keep the filled ground from settling and allowing the concrete to drop. This can be accomplished by fastening the edge of the concrete to the foundation, which is done by embedding ½-inch concrete reinforcing rods in the foundation. The reinforcing rods can be put in the foundation when it is poured or when the block is laid. (See *Foundation and Basement.*) Be sure that the outside edge of the concrete bears on firm, undisturbed soil.

Garage and basement floors are prepared similarly but there is no need for forms, except by the garage doors. Use a chalk line to mark the depth for the concrete on the walls. Basement floors must have a 4- to 6-inch gravel base; this is done by filling the area with gravel up to the top of the footing. Most gravel dealers will be able to put the gravel through a basement window. No reinforcing mesh is required for the basement, but it won't hurt if you want to be extra sure there are no cracks. The garage floor should have a 4-inch gravel base. Make forms by the doors on the outside edge of the wall, so the floor will extend under the doors. The garage floor should be about ½ inch above the driveway and slope up to 1 to 2 inches to the back of the garage. Use 6- x 6-inch reinforcing wire for the garage floor—this supports the edges where the fill around the footing will settle.

Use ready mix cement if it is available. Tell the supplier what you want it for so you will get the correct mix, and tell him how many cubic feet you need. Multiply the length and width (both in feet) of the section to be poured and multiply by the thickness in inches; then divide by 324. This will give you the cubic yards. Example: a driveway 9 feet by 18 feet by 4 inches thick, multiplied, equals 648. Then, 648 divided by 324 equals 2 cubic yards. Have plenty of help available when the cement arrives. This is at least a two-man job, and for large areas, three or four men will be better.

If ready mix cement is not available, you can mix your own. A standard mix is 1 part cement, 2 parts sand, and 3 parts gravel. The cement can be mixed by hand, but for jobs this big it is better to rent a gasoline- or electric-powered mixer. Mix the cement and sand first, then add the gravel. Add water until the mix is the right consistency, being careful not to use too much water.

Shovel the cement around until the surface is fairly level and the forms are full to overflowing. Screed it off level with the forms with a straight-edged board—have a man on each end of the board and saw it back and forth. There will now be a thin layer of water on the surface. As this water disappears, smooth the surface with a wood float. For driveways, walks, and patios, use an edging tool to form a rounded edge. No grooves are required if reinforcing is used. To provide a slip-resistant surface, outdoor concrete surfaces should be brushed by dragging a stiff brush across the concrete before it sets fully. Use straight, even strokes with a stiff-bristled push broom to get a pleasing texture.

Garage and basement floors should be steel-troweled to achieve a very smooth finish. This is a long job and requires going over the surface many times for about an eight-hour period, until the surface is hard. Watch an experienced cement finisher do this before you try it. If you have the basement or slab done professionally, you may learn enough by watching to do the garage yourself. A gas-powered rotary steel trowel can be rented for this job, and only the edges will have to be done by hand.

44. Brick

Brick is one building material which has changed considerably in recent years. Ten years ago, most houses were faced with smooth, even-colored face brick which was laid in neat, trim rows. The trend now is to use common building brick rather than face brick.

Odd-colored and varied-colored brick are now the best sellers —clinker or reject bricks which were once thrown away are now bringing premium prices. These twisted or warped bricks must be laid in slightly uneven rows; it gives a soft, pleasing texture to the home and is very much desired in the most expensive homes. It looks very good with the rough-sawn siding now used. (See *Exterior Siding and Trim.*)

An amateur bricklayer can lay common building brick in a manner which gives very professional results if no attempt is made to duplicate the more formal face brick style. If you prefer a more formal face brick laid straight and true, it would be wise to hire a bricklayer, as this takes considerable skill.

If you plan to have brick on one or more wall, the foundation must be built for this. Extend the foundation 5 inches outside the exterior dimensions of the walls. See the foundation drawings.

Before starting to lay bricks, figure out exactly how the bricks will be arranged. One brick plus the ⅜-inch mortar joints measures 8⅜ inches long by 2⅝ high. Lay the bricks out on the foundation and see where the windows and doors will fall. Plan the first course so that as few bricks as possible will have to be cut; adjust the dimensions by using larger or smaller mortar joints. Cut the bricks with a cold chisel. Stretch a plumb line from the eaves to the foundation on each end of the wall, and use this string to keep the corners straight. Locate the line so that the brick will be 1 inch from the sheathing. If the adjoining wall will have siding, extend the end bricks just far enough to cover the edge of the siding. It is best to have all trim and siding in place before laying the brick, so that the brick can be mortared up flush to the adjoining material. Also, figure the height of the bricks so they will come out even below the window sills and over the doors or windows. If the brick will go over the doors or windows, a steel lintel is required to hold it up. It is usually made of a 3½ x 3½ x ⅜-inch steel angle and should overlap the opening by 4 inches on each side. It is placed in the wet mortar with the flange extending up behind the next course of brick.

Provision must be made for openings in the brick; crawl space vents must be provided. (See *Flashings, Vents, and Gutters.*) The vents are laid in the brick wall to line up with matching holes in the floor header. Electric outlets and outside faucets must be cemented into the brick wall, as well. Use regular electric boxes and leave about a foot of wire hanging out. If a clothes dryer in the basement is vented through the floor header and the vent must be cemented into the wall, this must be done when the brick is laid. In some areas, the gas, electric, or water meters have windows so they can be read from the outside, and a glass block is installed in the brick wall for this. See the utility companies for instructions on how to do this.

The easiest mortar to use for laying brick is 1 part Portland cement, 1 part hydrated lime, and 6 parts sand. If you want to color the mortar, buy the coloring from your brick dealer, and get instructions on how much to use. Wet the bricks thoroughly before starting. Lay a ½-inch-thick layer of mortar on the top of the foundation, starting at the highest point if the foundation is not perfectly level. Stretch a string between two bricks about 12 feet apart and lay the first course to this string. Move the string as you go, and use a level to make sure the bricks are laid fairly straight. Always keep the corners or ends three or four courses higher than the rest of the wall and use the string to keep the corners level.

Spread the mortar over a large enough area to place two bricks at a time, then make a shallow furrow down the center of the mortar bed. Put a good dab of mortar on the end of a brick and press it into place against the previous brick. Cut off excess mortar flush with the brick with the edge of the trowel. The last brick in a course is laid by putting mortar on both ends. Never shift a brick once it is in position; if it must be moved, remove it and place new mortar. Every sixth course, nail 22-gauge metal brick ties to the wall studs and embed in the mortar joint. Space the ties 3 feet apart. Also, place ties within 12 inches of all window and door openings.

To finish the joints, run a round object such as a piece of pipe along the joints, pressing the mortar into the joints and rubbing off any excess. If you want raked joints, which are dug out, use a special joint tool and remove the mortar to a depth of ⅜ inch. Clean the brick thoroughly with a stiff brush and water after the mortar is set, but not dry. About a week after the bricks are laid, go over them again with a brush and a mixture of 1 part hydrochloric acid and 10 parts water. Rinse thoroughly with a garden hose to remove the acid.

BRICK DETAILS

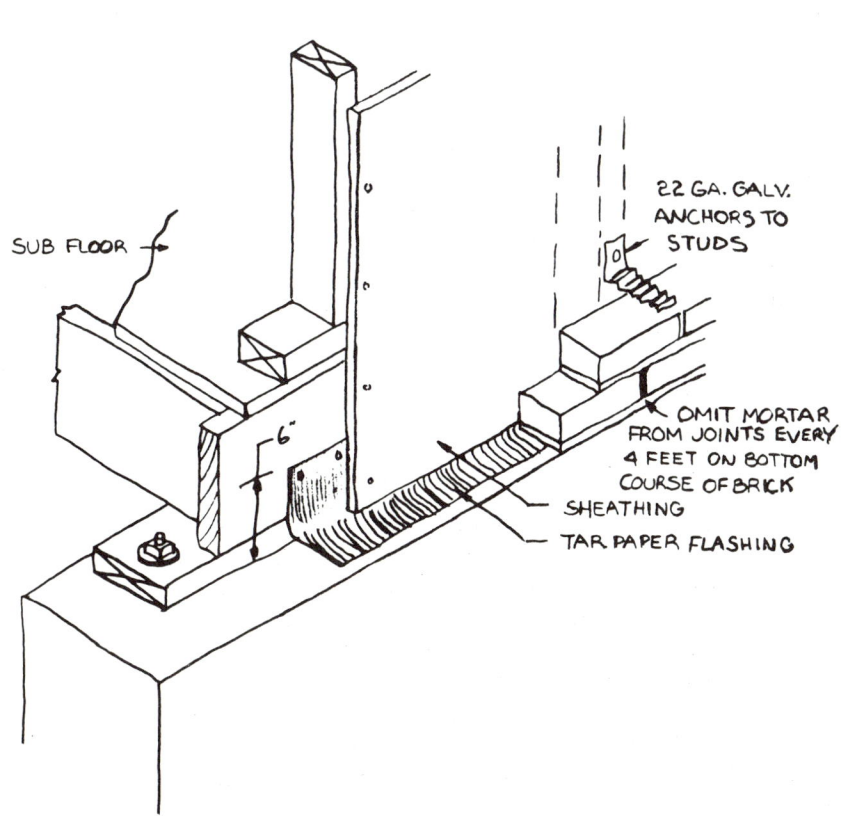

45. Fireplace

Fireplaces can be divided into two main types. One is used to heat a room or whole house. The other type is used as a decoration or novelty and the less heat it provides, the better, because it only interferes with the heating or air conditioning systems. Many homes now use the second type, but some vacation houses or houses in climates where furnaces are not used need a real fireplace. If you want a real brick or stone fireplace, a bricklayer who is experienced in fireplace design can help you plan for what you want. You will probably prefer to do it yourself, however, and prefabricated fireplace units are available in many types.

Brick or stone fireplaces can be built easily by buying a steel fireplace liner and building the fireplace around it. There are many stores and lumber dealers who sell these units and have information on how to install them. A fireplace of this type must be planned from the very start of the house. A 6-inch-thick footing must be provided under the fireplace area, and a concrete block foundation built up level with the floor. If you have a basement, a basement fireplace can be built in the same area to make double use of this space. The floor framing is altered by building headers around the fireplace area. The fireplace continues up through the ceiling joists, which are also cut and headers installed. In the attic area, only the chimney portion needs to continue through, unless you want a large chimney for reasons of appearance. Cut the roof rafters to clear the chimney and install headers.

A much easier fireplace to install is the zero-clearance type. It is much lower in cost and it takes less space, but can burn wood logs and have a brick face just like a real brick fireplace. This is probably the best bet for the do-it-yourself builder, and is used by many building contractors. The fireplace is purchased knocked down and is assembled on the job. It can be placed directly on the subfloor, although if you will have a brick front, the floor joists should be tripled under the area that will support the brick. If the fireplace is at right angles to the floor joists, double all the joists under the fireplace area. A false wall can be built flush with the front of the fireplace to give a built-in look. If the house is on a slab, a whole brick fireplace wall can be assembled and put in place on the fireplace. A hole is cut through the ceiling and roof. If any joists or rafters are cut, nail in a header. A prefabricated chimney housing is used to complete the job.

Gas and electric fireplaces are also available and can be put in place after the house is done. Some types are free standing and others are fastened to the wall. The gas models require an easily installed chimney flue and a gas connection. The electric models require either a 115- or 230-volt electric circuit, but no chimney.

46. Grading, Seeding, and Sodding

After the walks, driveway, and patio are done and all the utility lines are in, the lawn can be graded. The best time to plant a lawn is in September when the weather is cooler and the weeds have stopped growing. Lawns planted in the spring have to battle with fast-growing weeds. The soil is also easier to work in the fall.

Remove any weeds which have grown during the construction. You can hire someone with a tractor to do the grading, or rent a tractor and do it yourself. It should be graded to let water run away from the house, and a low spot or swale should be left between houses so water from one yard will not run onto another. The lot should be graded as evenly as possible and then pulverized.

If you seed the yard, use a mixture which is at least 75 per cent Kentucky bluegrass, which is good for most of the country. Avoid mixtures with high percentages of rye or Timothy. Sow the seed evenly over the ground—use a spreader to get an even coverage—and keep the seed moist until it starts to sprout. It may take three or four weeks until the lawn is good and green. About three months later, or in the following spring, spread fertilizer over the lawn—this should also be done with a spreader. Follow the fertilizer manufacturer's directions.

If you prefer to sod the lot, you can buy sod at a cash-and-carry price from most sod farms. The best time of year to lay sod is in the spring. Lay the sod immediately after it is purchased, or buy it late in the afternoon and lay it early in the morning. Don't try to lay sod in hot sunny weather because it will dry out and die. Prepare the soil the same as for seeding, then lay the strips close together and fill the cracks with fine soil. Tamp the sod to get it in firm contact with the soil. Keep the lawn well watered until it is established.

When you plant trees, don't plant over areas which were excavated for utility lines, because these areas have a tendency to settle for as many as five years. The grass planted on areas which have settled can be removed with a flat spade or shovel and the area can be filled with dirt to the level of the rest of the lawn. The grass can then be replanted like sod. Also, in the rare case that utility lines have to be dug up, you will not want to remove any trees.

47. Lumber Sizes

Product Standard 20-70 promulgated by the U.S. Department of Commerce and in effect January, 1971, establishes a relationship between unseasoned and dry lumber sizes. Minimum standard-surfaced sizes at the time of manufacture for dry lumber are shown below.

Nominal Size	New Sizes	Old Sizes
1 x 2	3/4 x 1-1/2	25/32 x 1-5/8
1 x 3	3/4 x 2-1/2	25/32 x 2-5/8
1 x 4	3/4 x 3-1/2	25/32 x 3-1/2
1 x 6	3/4 x 5-1/2	25/32 x 5-1/2
1 x 8	3/4 x 7-1/4	25/32 x 7-1/4
1 x 10	3/4 x 9-1/4	25/32 x 9-1/4
1 x 12	3/4 x 11-1/4	25/32 x 11-1/4
2 x 2	1-1/2 x 1-1/2	1-5/8 x 1-5/8
2 x 4	1-1/2 x 3-1/2	1-5/8 x 3-5/8
2 x 6	1-1/2 x 5-1/2	1-5/8 x 5-1/2
2 x 8	1-1/2 x 7-1/4	1-5/8 x 7-1/2
2 x 10	1-1/2 x 9-1/4	1-5/8 x 9-1/2
2 x 12	1-1/2 x 11-1/4	1-5/8 x 11-1/2
4 x 4	3-1/2 x 3-1/2	3-5/8 x 3-5/8
4 x 8	3-1/2 x 7-1/4	3-5/8 x 7-1/2
4 x 10	3-1/2 x 11-1/4	3-5/8 x 11-1/2

Product Standard 20–70 defines dry lumber as having a moisture content of 19 per cent or less and unseasoned lumber as having a moisture content of over 19 per cent.

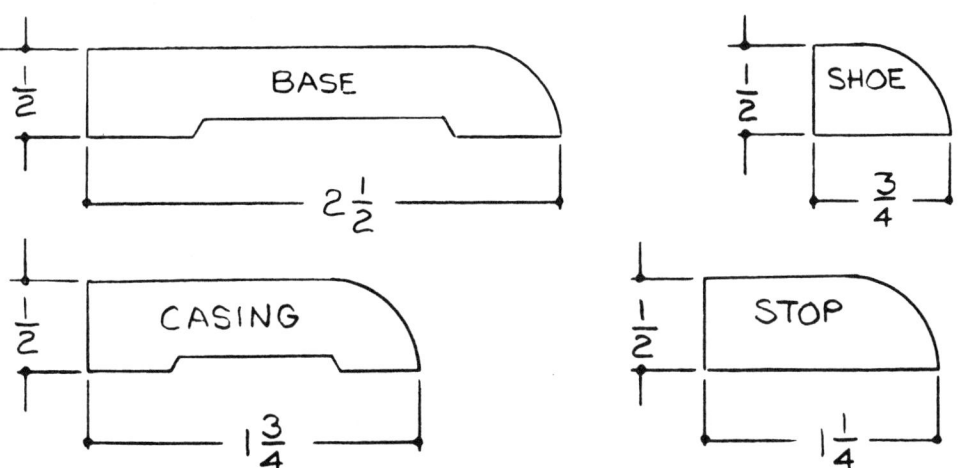

TYPICAL TRIM DETAILS

48. Nails and Screws

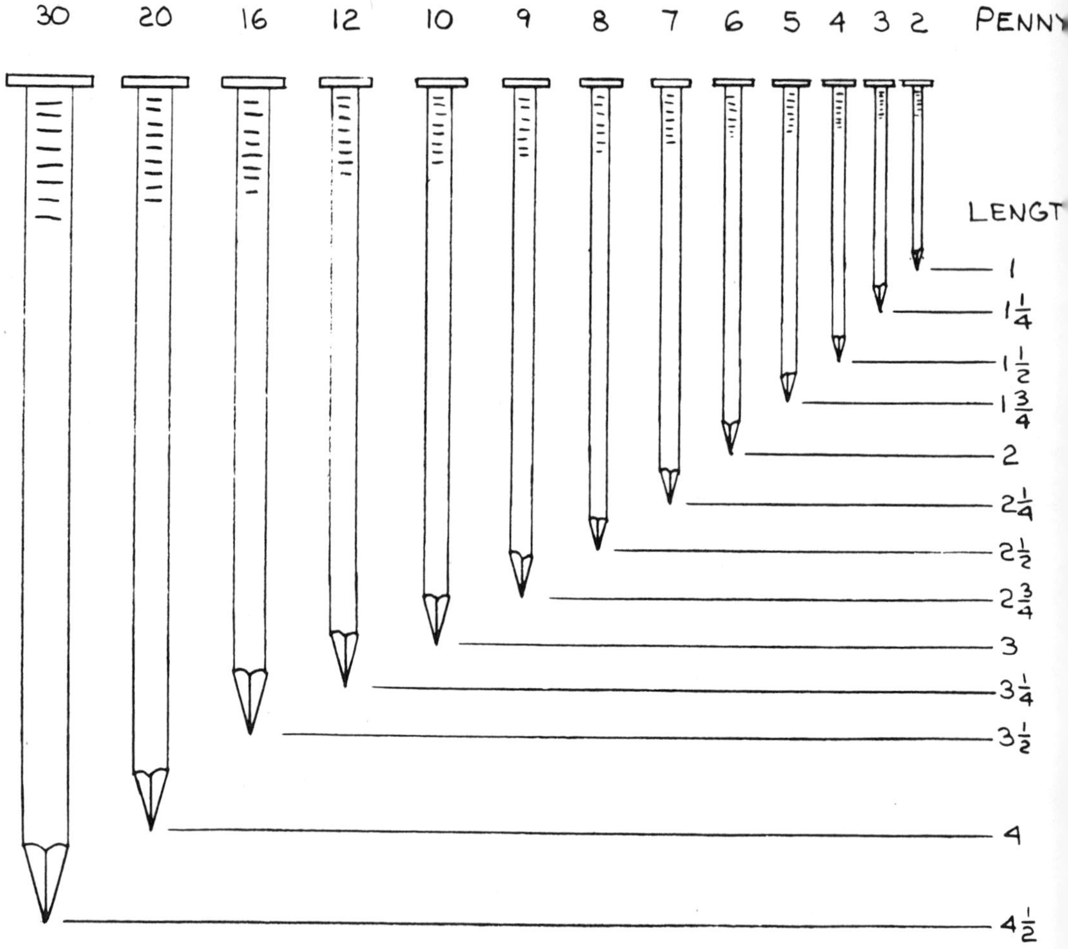

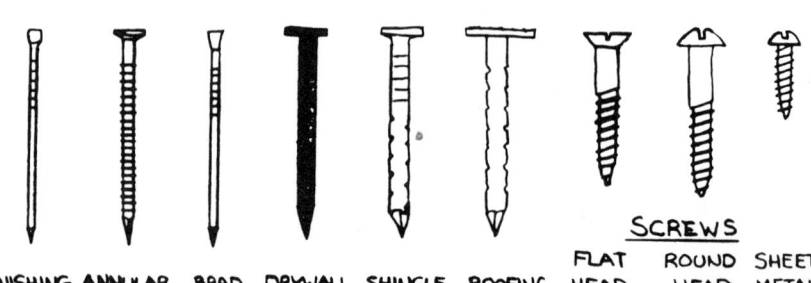

49. Tools

The following list of tools you may need is furnished as a guide only. You may already own many of them, and some you will want to buy. Others may be rented.

Description	Cost
Auger, post hole	$ 7
Broom, push	4
Caulking gun	2
Chalk line	2
Chisel, cold—3-piece set	2
Chisel, wood—1 inch	3
Conduit bender—1/2- and 3/4-in.	3
Cord, electrical—100 feet	13
Drill, electric hand—3/8-in.	25
Drill set, masonry—4-piece	5
Drill set, metal—13-piece	4
Drill set, wood—5-piece	5
Drywall knife	2
File, mill—10-inch	1
Float, cement finishing	5
Garden hose—100 feet	14
Hacksaw	3
Hammer, bricklayer's	5
Hammer, carpenter's	5
Hammer, sledge—8-pound	9
Hatchet, carpenter's	7
Level, line	1
Ladder, extension—16-foot	30
Ladder, step—6-foot aluminum	20
Level, aluminum—24-inch	7

182 TOOLS

Miter box, or	1
Miter box, steel with saw	18
Paint brushes, 4-, 3-, 2-inch	10
Paint brushes, sash—1-1/2-, 2-in.	4
Paint roller set	5
Pipe threading set—1/2-, 1-inch	36
Pipe wrench—14-inch	5
Pipe wrench—24-inch	11
Plane, hand	9
Pliers, are joint—12-inch	4
Pliers, ceramic tile	4
Pliers, lineman's—7-inch	4
Pliers, regular—6-inch	2
Plumb bob, steel	1
Putty knife	1
Rule, folding—6-foot	2
Sander, power	20
Saw, coping	1
Saw, electric hand	35
Saw, hand—26-inch crosscut	8
Saw, saber	30
Saw, table or radial arm	250
Screw drivers—6-piece set	6
Shears, sheet metal—10-inch	3
Shovel, D-grip	5
Shovel, straight handle	5
Square, carpenter's	4
Stapler	11
Stud driver	4
Tape, steel—10-foot	2
Tape, steel—100-foot	8
Torch kit, propane	10
Trowel, brick	3
Tubing cutter	4
Twine, nylon—100 yards	1
Vise, pipe—1/8, 1-1/2-inch	7
Vise grips—10-inch	2
Wheel barrow, garden type	18
Wire strippers	4
Wrecking bar—30-inch	3
Wrench, adjustable—8-inch	3
Wrench, adjustable—12-inch	6

50. Glossary of Terms

ACOUSTICAL TILE—soundproofing ceiling tiles
ADAPTER—pipe fitting used to connect two different types of pipe
AGGREGATE—stone used in concrete
AMPERE or AMP.—measurement of electrical current or flow
ANCHOR BOLTS—bolts in the foundation which hold the sill in place
ARMORED CABLE—electrical cable with a flexible steel cover
ASPHALT SHINGLES—tar paper or composition roofing
ATTIC—unfinished space between the ceiling and roof
BACKFILL—to refill excavations with earth
BASE—trim used against the wall, next to the floor
BASE CABINETS—kitchen cabinets below the counter top
BASEMENT—an area more than half below ground level
BATT INSULATION—insulation which comes in 14-1/2-inch-wide rolls
BATTENS—narrow boards to cover vertical joints in siding
BATTER BOARDS—corner boards used to hold strings to lay out building site
BAY WINDOW—window which projects out from the wall
BEAM—heavy wood or steel member to support parts of a house
BEARING PARTITION—interior partition which supports the roof, ceiling or second story
BEARING TEST—test of soil to determine the load it will support
BIFOLD DOORS—doors with hinge in the middle which fold completely out of the way, used for closets
BLACKTOP—asphalt-type driveway topping
BLOCK—see CONCRETE BLOCK
BLOCKING—boards used between framing members to provide nailing surface

184 GLOSSARY OF TERMS

BONDING—gluing

BOOT—furnace fitting used to connect ducts to register box

BOSTON RIDGE—method of finishing the ridge of shingle roofs

BOX—metal holder for electrical switch or receptacle

BRACING—diagonal strengthening

BRICK VENEER—brick covering used on frame houses

BRIDGING—cross bracing between floor joists

BTU—British thermal unit—measurement of heat, used in heating and air conditioning

BUILT-UP ROOF—roof built up with three to five layers of asphalt felt

BUILDING DRAIN—sewer pipe from the soil stack to sewer or septic tank

BUTT—a joint formed by two pieces of wood joined end to end

BX—see ARMORED CABLE

CAP—pipe fitting used to close the end of a pipe

CARPORT—roofed space for cars with open sides

CASEMENT—a window with hinges on the vertical edges

CASING—trim for around window and door frames

CAULKING—putty for sealing joints

CAVITY—holes in blocks, or void space

CEMENT—substance mixed with sand and water to make mortar or concrete; also, a type of glue

CERAMIC TILE—glazed finishing tiles used in bathrooms

CESSPOOL—tank or pit used to let rain water soak into the ground

CHALK LINE—a chalk-covered string used to mark a straight line

CHIMNEY—a pipe to exhaust fumes from the furnace, fireplace or water heater

CIRCUIT—complete, working electrical system

CIRCUIT BREAKER—automatic safety switch, replaces fuses

CLAPBOARD—barn-type overlapping exterior siding

CLOSET—closable storage area within a house

CLOSET FLOOR FLANGE—connection between the toilet and drain pipe

COLLAR BEAM— connects opposite roof rafters to prevent lifting

COLUMN—post which supports part of the house

COMBINATION—window or door with interchangeable or movable screens and storm sash

COMMON BRICK—building brick once used only for structural purposes
CONCRETE—mixture of sand, gravel, cement, and water
CONCRETE BLOCKS—hollow blocks made of concrete, used for foundations
CONCRETE FLATWORK—floors, walks, driveways, etc.
CONDENSER—part of an air conditioner which exhausts heat
CONDUCTOR—a single electrical wire
CONDUIT—tubing used to contain electrical wires
CONNECTING WASTE—drain fitting used to connect kitchen sink drain traps
CONNECTOR—fitting used to connect wire or conduit to electrical boxes
COPING SAW—a thin-bladed saw used to cut contours in trim
CORNER BEAD—wood or metal molding used on outside corners
COUNTER TOPS—plastic laminated top for kitchen cabinets
COUPLING—pipe fitting used to permanently connect two pipes; also see UNION
COURSE—one horizontal row of brick, block or shingles
COVE—wood or metal molding used on inside corners
CURRENT—the flow of electricity
d—see PENNY
DAMPER—air control door used in fireplaces and heat pipes
DECK—an outdoor wood floor or walkway
DEFLECTION—bending of a framing member
DETAILS—drawings showing methods of construction
DISPOSAL FIELD—an array of tile used with septic systems
DORMER—a window which projects out from the roof
DOUBLE HUNG—common windows with two vertically sliding sashes
DOWNSPOUTS—pipes to carry water down from gutters
DRAINPIPE—sewer pipe
DRAIN TRAP—a water-filled trap at each plumbing fixture drain to prevent exit of gases
DRIP CAP—exterior trim over doors and windows
DRYWALL—gypsum board with paper surface on both sides used in place of plaster
DUCTS—heating or air conditioning pipes

EAVES—roof overhang which extends out from the house

ELBOW or ELL—pipe fitting to join two pipes at 90° or 45° angle

ELEVATIONS—drawings of side views of a house

ENTRANCE ELL—an electrical conduit fitting which has a removable cover for pulling wires

EVAPORATOR—part of an air conditioner which removes heat from the air

EXPANSION JOINT—space left between materials to allow for expansion caused by heat or moisture

EXTERIOR—any part of the house exposed to weather

FACE BRICK—formal brick used for exposed surfaces

FACE NAIL—nails driven straight through one board into another

FASCIA—board which closes the end of the rafters

FAUCET—value to allow water to leave pipes

FHA—Federal Housing Administration—government body which sets building standards for government loans

FELT—asphalt-saturated material used under shingles or for waterproofing

FIBERGLASS—fibers made from glass and used for insulation

FILL—earth or gravel used to fill voids around the foundation or concrete

FINISHED GRADE—the level of the earth around the finished house

FINISHING NAILS—nails with very small heads used for trim

FIR—wood commonly used for plywood and framing

FITTINGS—see PIPE FITTINGS

FIXTURES—tubs, lavatories, toilets, showers; also ceiling-mounted lighting devices

FLASHING—sheet metal used for protection against water

FLOOR COVERING—finish floor material: tile, linoleum, carpet

FLOOR PLAN—a drawing showing the top view of the floor and walls of a house

FLOORING—the finish material used for floors

FLUES—pipes or linings used in chimneys

FLUX—compound for cleaning metal to be soldered

FOOTINGS—concrete base on which foundation is built

FORMS—boarding used to contain and give a shape to wet concrete

FOUNDATION—masonry walls which support a house

GLOSSARY OF TERMS 187

FRAMING—wooden skeleton of a house

FROST LINE—line below which soil does not freeze

FURRING—narrow wood strips used to space ceiling tile or plywood from wall or ceilings

GABLE ROOF—pitched roof with two sides

GABLE—peaked end of a gable roof

GALVANIZED STEEL—sheet steel with zinc coating

GAMBREL ROOF—roof commonly used on barns; provides additional attic space in houses

GARAGE—closed-in area for cars

GAUGE—thickness of metal or wire

GRADE—see FINISHED GRADE

GROUT—masonry used to fill narrow cavities

GUTTER—trough used to catch water running off the roof

GYPSUM BOARD—gypsum covered with paper used in drywall construction

HALF-BATH—a bath having only a toilet and lavatory

HANGERS—steel supports used in framing; also, supports for electrical boxes and furnace ducts

HARDBOARD—a pressed-wood composition board used for floor underlayment

HEADERS—lumber running at a right angle to joists used to hold up ends of cut joists; lumber used to close ends of joists flush with outside walls; also, framing over doors and windows

HIP ROOF—pitched roof having four sides

HOLLOW-CORE DOOR—an interior door with plywood surfaces and hollow frame

HOOD—sheet metal collector for smoke, used over the stove, oven or fireplace

I-BEAM—steel beam used to support the house

INSULATION—material used to prevent heat loss or gain

INSULATION BOARD—exterior sheathing with high insulation value

INTERIOR—any part of the house not exposed to weather

JAMBS—vertical sections of the door and window frames

JOISTS—horizontal framing members which support the floors or ceilings

LAG BOLT—large wood screw with a square head

LAVATORY—bathroom washbowl

LINTELS—wood or metal framing over the doors, windows and arches

LIVING AREA—the floor area of a house not including basement, garage or exterior storage

LOUVERS—narrow boards spaced and angled to stop vision but allowing air to flow through

MASONITE—see HARDBOARD

MEMBRANE—plastic sheet used for waterproofing

METER SOCKET—box which holds electrical meter

MILLWORK—windows, door, wood moldings, etc.

MITER JOINT—joint made by two boards each cut at a 45° angle and joined at a right angle

MORTAR—cement used for bricks or blocks

MOLDING—wood milled to a shape for trim

MUD ROOM—a half-bath near rear door, garage or utility room

MULLION—vertical member between two windows set in one opening

NIPPLE—short piece of pipe with threads on each end

NOMINAL LUMBER—the state rather than actual lumber size, as in 2 x 4

NONMETALLIC-SHEATHED CABLE—electrical cable with plastic covering

OUTLETS—exit points for heat or air conditioning; also see RECEPTACLES

OVERHANG—roof eave which extends out from the house

P-TRAP—a type of drain trap

PANELING—wood or hardboard sheets used to cover walls, usually prefinished

PARGING—a coat of plasterlike cement to cover a foundation

PARTITIONS—interior walls which separate rooms

PEA GRAVEL—a clean, smooth pea-sized gravel used in crawl space

PENNY or d—measurement of nails by weight

PHILLIPS SCREW—screw with cross-shaped slot in head

PIERS—masonry posts used to support buildings

PINE—lumber commonly used for doors, windows, and trim

PIPE—tubes used to carry gas, water, sewage

PIPE FITTING—part used to join pipes and fixtures

PITCH—angle of roof or pipes expressed in feet of rise in 12 feet, or inches per foot

PLAN—a drawing showing a top view of house
PLASTER—interior wall covering applied with a trowel
PLATE—frame member used at top and bottom of a wall
PLATFORM—commonly used method of framing a house, also called Western framing
PLENUM—distribution or collection ducts for a furnace
PLOT—parcel of land
PLUG—pipe fitting used to stop an opening
PLUMB BOB—a weight used on the end of a string to determine a vertical line
PLYWOOD—sheets of lumber made by gluing thin sheets together for greater strength
POCKET DOOR—door which slides into a pocket in the wall
POWDER ROOM—bath having only toilet and lavatory
RAFTERS—framing members used to support the roof
RAKED JOINTS—brick walls with mortar raked out of the joints to provide shadows
RANGE—kitchen stove
RECEPTACLE—electrical outlet
REDUCER—pipe fitting for joining two pipes of different sizes
REGISTER BOX—furnace fitting to connect the boot to the heat or air conditioning outlet
REINFORCED CONCRETE—concrete containing wire mesh
RIDGE—framing member which forms peak of roof
RISER—vertical part of the stair step
ROMEX—see NONMETALLIC-SHEATHED CABLE
ROOFING—top covering of the roof
ROUGH-IN—partial plumbing or electrical system to allow walls and ceilings to be finished
ROUGH OPENING—oversized opening for installation of a door or window
SANITARY TEE—special T-shaped fitting used in drainage plumbing
SASH—movable part of the window
SCREED—a board used to level the surface of concrete
SEPTIC SYSTEM—sewage disposal system
SEPTIC TANK—collection and distribution pit for sewage
SERVICE ENTRANCE—apparatus to bring in electrical power

SERVICE ENTRANCE PANEL—electrical fuse box
SERVICE HEAD—entrance for electrical wires
SHAKES—wooden roof shingles
SHEATHING—covering for exterior walls and roof used under the siding or roofing
SHED ROOF—pitched roof having only one side
SHEET ROCK—gypsum board or drywall
SHIM—a thin wood strip or shingle used as a spacer
SHINGLES—overlapping wood or asphalt covering for roof or exterior walls
SHIPLAP—an overlapping joint in horizontal siding
SHOE—molding used between the base trim and finished floor
SIDING—outer covering for exterior walls
SILL—framing member bolted to the foundation; also, lower horizontal part of the door and window frames
SITE—location for a house
SKYLIGHT—a window in the roof, usually made of translucent plastic
SLAB—concrete floor
SLIDING DOORS—usually aluminum-framed glass doors with one half fixed and one half sliding
SOFFITS—lowered ceiling over the kitchen cabinets
SOIL PIPE—sewer pipe
SOIL STACK—sewer and vent pipe from building drain to roof
SOLE PLATE—the framing member at the base of walls and partitions
SPAN—the unsupported length of a joist or rafter
STEPPED FOOTING—footing with a step on the outside to allow brick to extend to the ground; also, footing with a step on the inside to allow the floor joists to sit lower
STOOL—protruding trim at interior windowsill
STOP—trim strip in the door frame which the door stops against
STOP-AND-WASTE VALVE—water supply valve with small drain fitting
STORM DOOR—wood or aluminum exterior door with interchangeable glass and screen
STORM WINDOW—aluminum-framed window with sliding or removable glass and screen
STORM SEWER—special sewer for catching rain water
STUDS—vertical framing members used in the walls

SUBFLOOR—the rough floor laid on the floor joists

SUMP PUMP—pump used to pump water from the basement

SUPPLY—water pipe

SUSPENDED CEILING—a ceiling finish consisting of a metal frame with removable panels

SWITCHES—wall switch to control lights or outlets

SWIVEL TRAP—drain trap used with bathtubs

TAKE-OFF—furnace fitting to connect the air ducts to the plenum

TEE—T-shaped pipe fitting

TERMITE SHIELD—a metal shield on top of the foundation to protect against termites

THERMOPANE—double glass with air space, used in place of storm windows

THERMOSTAT—electric switch which responds to temperature

THRESHOLD—aluminum and plastic cover for exterior door sills

TIES—metal straps to hold brick to wall

TILE—ceramic floor or wall tile; acoustical ceiling tile; vinyl or other floor tile; carpet tile; also, sewer tile or drain tile for draining water from around footings

TOENAIL or TN—a nail driven at an angle to fasten two boards

TOGGLE BOLTS—bolts which are inserted in a hole in a wall and then open inside the wall, commonly used in drywall

TONGUE AND GROOVE—lumber with interlocking parts

TOP PLATE—the two framing members at the tops of walls and partitions

TRAPS—see DRAIN TRAP

TREAD—flat part of the stair step

TRIM—wood around windows and doors and where walls meet floors

UNDERLAYMENT—plywood or hardboard floor covering to be used under tile or carpet

UNION—pipe fitting used to allow disjoining of pipes

UTILITIES—gas, water, electricity, sewer, telephone

UTILITY ROOM—a room used for the furnace, water heater, washer and dryer, etc.

VALLEY—low point between intersecting roofs

VALVE—device to shut off water or gas supply

VANITY—a cabinet below the lavatory

VAPOR BARRIER—plastic membrane used to seal out moisture

VENTS—louvered or screened openings in crawl space, eaves or roof to ventilate area; also, return pipes from plumbing drains to the soil stack

VERTICAL SIDING—siding applied vertically

VOLTAGE or VOLT—measurement of electrical pressure

WATER TABLE—height of underground water

WATT—the unit of electrical power

WEATHER STRIPPING—metal or fiber strip used around doors and windows to keep out wind, rain, etc.

WEEP HOLES—holes in brick wall at the bottom to let moisture out

WIRING—electrical wiring

Y-BRANCH—sewer pipe with side outlet at 45° angle

51. Specifications

Contents	Section
General Instructions	A
Concrete and Masonry	B
Carpentry and Millwork	C
Floor Finish	D
Wall Finish	E
Plumbing	F
Electrical	G
Heating and Air Conditioning	H
Sheet Metal	I

Section A: General Instructions

Instructions: This outline of specifications, when filled out by the owner-builder, will set forth the specifications for the residence. Cross out the specifications which do not apply. In case of any conflicts between these specifications and the building plans, the specifications will be the authority. If some details shown on the plans are not covered by these specifications, the plans will be the authority.

Attorney: An attorney will be consulted before any contracts are signed or agreements made.

Building Codes: The owner-builder and any subcontractors will comply with all building codes and ordinances.

Insurance: The owner-builder will provide fire, wind, and liability insurance during the entire period of construction.

Section B: Concrete and Masonry

Excavation: The excavation for footings will be _____ inches below the finished grade.

Footings: The footings for frame walls will be 6 inches thick and _____ inches wide. The footings for brick veneer walls will be 6 inches thick and _____ inches wide.

Foundation: The foundation for frame walls will be _____ inches thick. The foundation for brick veneer walls will be _____ inches thick. The foundation material will be (poured concrete) (concrete block).

Basement floor: The basement floor will be 4 inches thick and have a 4- to 6-inch gravel base.

Crawl space floor: The floor will have a plastic membrane vapor barrier with a covering of pea gravel.

Slab floor: Slab floors will have a 4-inch gravel base, a plastic membrane vapor barrier, and perimeter insulation. The slab will be 4 inches thick and have a 6- x 6-inch wire mesh reinforcing.

Garage floors: Garage floors will have a 4-inch gravel base and a 4-inch-thick slab with a 6- x 6-inch wire mesh reinforcing.

Driveway, walks, and patio: All exterior concrete work will be reinforced with a 6- x 6-inch wire mesh. The concrete will be 4 inches thick. The driveway will be _____ feet wide. The front walk will be 3 feet wide. The patio will be _____ by _____ feet.

Brick and stone: (Brick) (Stone) will be used as follows: (front elevation) (rear elevation) (left elevation) (right elevation). Height of brick will be: (to window sills) (to eaves). All sills or tops of partial brick or stone veneer will be capped with sloped (rowlock brick) (natural cut stone). (Brick) (Stone) will be: name _____ color _____ . Brickwork to be thoroughly scrubbed with 10% muriatic acid. Steel lintels will be provided over all openings.

Joints: Joists will be type _____ color _____ .

Fireplace: Fireplace will be constructed as required by plans.

Section C: Carpentry and Millwork

Framing lumber: Framing will be species: _____ .
Floor girder: (3- 2 x 10 framing lumber) (7-inch steel I@ 15.3#)
Floor joists: (2 x 8) (2 x 10) spaced 16 inches on centers
Studdings: 2 x 4 spaced 16 inches on centers

Bridging: 1 x 3
Subfloor: 5/8-inch exterior plywood
Ceiling joists: 2 x 6 spaced 16 inches on centers
Roof rafters: 2 x 6 spaced 16 inches on centers
Collar ties: 2 x 4 spaced 4 feet on centers
Wall sheathing: (3/8) (1/2) exterior plywood
Roof sheathing: 3/8-inch exterior plywood
Roofing: (Wood shingles) (Asphalt shingles) _____ brand _____ weight or size _____ color _____
Exterior siding: Style _____ species or material _____ and will be applied (horizontally) (vertically)
Exterior trim: Species or material _____
Exterior doors: Style _____ species _____
Storm doors: Style _____ species or material _____
Sliding glass doors: Manufacturer _____ style number _____
Garage door: Manufacturer _____ style number _____
Windows: Manufacturer _____ style number _____
Basement windows: Steel frame
Interior trim: Style _____ species _____
Interior doors: Style _____ species _____
Closet doors: (Sliding) (Bifold) style _____ species or material _____
Kitchen and bath cabinets: (Custom-made) (Manufactured) material _____ manufacturer _____ style number _____
Counter tops: Pressure-laminated plastic, color _____ style number _____
Stairs: Treads will be 1-1/8-inches thick, species _____
Walls: Thickness _____ type _____
Ceiling: Thickness _____ type _____
Insulation: Floor slab thickness _____ width _____ type _____
Caulking: All exterior joints will be caulked.
Closets: All closets will have shelf and pole.

Section D: Floor Finish

Underlayment: Will be used under all but hardwood floor; 3/8-inch interior plywood will be used.
Finish flooring: Material _____ color _____ type _____
Living room:

Kitchen:
Dining room:
Entry hall:
Main hall:
Bedrooms:
Baths:
Other rooms:

Section E: Wall Finish

Wall material: Walls will be 1/2-inch gypsum wallboard.
Ceiling material: Ceilings will be (1/2) (3/8) -inch gypsum wallboard.
Other wall finish: Material _____ color _____ type _____
Living room:
Kitchen:
Dining room:
Entry hall:
Main hall:
Bedrooms:
Baths:
Garage:
Other Ceiling Finishes: Material _____ color _____ type _____
Living room:
Kitchen:
Dining room:
Entry hall:
Main hall:
Bedrooms:
Baths:
Garage:
Taping of seams: All gypsum wallboard used as a finish wall or ceiling will have the joints taped and filled by a skilled craftsman. All exterior corners will have metal corner beads.
Wall tile: Baths will have ceramic tile, color _____ type _____ , installed as shown on the plans.

Section F: Plumbing

Sewer from street to house: Will be steel or iron pipe.

Sewer under the house: Will be (cast-iron pipe) (hubless cast-iron) (plastic pipe).
Exposed sewer and vent pipes: Will be (hubless cast-iron) (plastic) (copper).
Cleanouts: Will be provided.
Cold water supply pipes: Will be 3/4-inch (copper) (galvanized pipe) (plastic). Branch pipes will be 1/2-inch.
Hot water pipes: Will be 1/2-inch (copper) (plastic) pipe.
Gas pipes: Will be black steel pipe.
Toilet: Type _____ color _____
Lavatory: Type_____ color _____
Bathtub: Steel or cast iron. Color_____
Kitchen sink: (Stainless steel) (Enamel). Type _____ color _____
Fixtures: Will be of (single) (double) faucet type with mechanical drains. Tub will have shower heads.
Garbage disposal: Manufacturer_____ model _____ color _____
Dishwasher: Manufacturer _____ model _____ color _____
Washer and dryer: Manufacturer_____ model _____ color _____
Laundry tub: Size _____ (single) (double)
Hot water heater: Will be (electric) (gas) (oil). Gallon capacity_____ recovery rate, gallons per hour _____ manufacturer _____
Water softener: Gallon capacity _____ manufacturer _____
Well and pump: Will be _____
Septic tank: Will be _____

Section G: Electrical

Electric service: Will be (100 amps) (200 amps) 115/230 volts.
Switches and receptacles: Will be located as shown on the plan.
Circuit breakers: Kitchen circuits will be 15-amp. Lighting circuits will be 10-amp.
Special outlets: Provided for: (Basement or garage power tools) (230-volt window air conditioner) (kitchen ventilator fan) (bathroom ventilator fan) (time-delay entrance lights) (intercom) (garbage disposal) (outdoor outlets) (outdoor lights) (electric garage door) (doorbell or chimes) (wall-hung clocks) (230-volt electric range) (230-volt electric dryer) (attic fan) (furnace) (air conditioning)
Lighting fixtures: Number _____ type _____
Range hood: Manufacturer _____ model _____

Attic fan: Manufacturer _____ model _____
Bath ventilator: Manufacturer _____ model _____
TV antenna system: Number of outlets_____ manufacturer _____ model _____
Electric garage door opener: Manufacturer_____ model _____
Attic fan: Manufacturer _____ model _____
Intercom: Manufacturer _____ model _____

Section H: Heating and Air Conditioning

Fuel: Will be (electricity) (gas) (oil) other _____.
Furnace: Will be (electric base boards) (hot water base boards) (forced warm air) other _____ furnace size _____ BTU _____ manufacturer _____ model_____
Extra features will be: (Humidifier) (air purifier)
Chimney: Will be (prefabricated metal flue) (masonry)
Air conditioning: Will be (gas) (electric), size_____ BTU _____ manufacturer _____ model _____

Section I: Sheet Metal

Termite shield: A termite shield of (copper) (aluminum) will be used under all wood sills.
Flashing: A flashing of (copper) (aluminum) will be used over all windows and doors. Also, flashing is used around all vents and the chimney in the roof.
Vents: Aluminum vent pipe will be used for kitchen and bathroom and laundry ventilators. Roof vents will be (plastic) (aluminum) and eave vents will be (aluminum) (galvanized).
Gutters: Will be box-type with rectangular downspouts. Material will be (steel) (aluminum) with (galvanized) (enamel) (natural) finish.

52. Materials List

This materials list is furnished as a guide to the materials you will need to build your home. By using this list, you can determine the cost of the materials required for a wide variety of houses. Quantities are omitted; they would only be confusing due to the many variations caused by personal preferences. In some cases, sizes of materials are omitted where the size varies with the design. The correct sizes for the materials you plan to use can be found in the instructions or on the plans.

Read over the materials list, and mark out or change any items which you do not want to use on the house you are building. Fill in the quantities and sizes of materials which are not given. Quantities of some materials, such as paint or nails, can be estimated by dealers who sell these every day. If you have a question on sizes or quantities of anything, show the plans to your materials dealer and he will be able to help you.

The main advantage of the materials list is to make sure you don't forget any costly items. This can be an unhappy experience if you are building with a fixed amount of cash to complete the job.

Materials—Type and Size

Footings
Cu. Yds.	Ready-mix concrete, or
Cu. Yds.	Sand
Cu. Yds.	Gravel
Sacks	Cement

Basement Foundation
Unit	Poured concrete foundation, or
Pcs.	12 x 8 x 16 masonry block (walls)
Pcs.	12 x 8 x 16 masonry corner block (corners)
Pcs.	8 x 8 x 16 masonry block (top course brick veneer)
Pcs.	8 x 8 x 16 masonry block (garage walls)

MATERIALS LIST

Pcs.	8 x 8 x 16 masonry corner block (garage corners)
Pcs.	6 x 8 x 16 masonry block (top course garage)
Pcs.	12 x 8 x 16 masonry window channel block
Pcs.	12 x 8 x 8 masonry window channel block
Cu. Yds.	Mason's sand
Sacks	Mortar
Sacks	Portland cement (waterproofing)
Gallons	Asphalt foundation coating (waterproofing)
Pcs.	16 x 32 steel sash basement windows
Pcs.	7-inch steel I-beam at 15.3 pounds x _____ lg.
Pcs.	8 x 8 x 3/8 steel bearing plates
Pcs.	3-inch steel pipe posts x 7 ft. lg.
Pcs.	4 x 4 x 1/4 steel shims
Pcs.	1/2 x 8 bolts with nuts and washers

Basement Floor

Cu. Yds.	Fill gravel
Cu. Yds.	Ready-mix concrete, or
Cu. Yds.	Sand
Cu. Yds.	Gravel
Sacks	Cement

Basement Foundation Tile

Pcs.	4-inch solid plastic pipe x 10 ft.
Pcs.	4-inch perforated plastic pipe x 10 ft.
Pcs.	4-inch 45° bends
Cu. Yds.	Crushed rock

Crawl Space, Slab, and Garage Foundation

Pcs.	8 x 8 x 16 masonry block (frame)
Pcs.	10 x 8 x 16 masonry block (brick veneer)
Pcs.	6 x 8 x 16 masonry block (top course brick veneer)
Pcs.	4 x 8 x 16 masonry block (top course slab-frame)
Pcs.	6 x 8 x 16 masonry block (garage)
Pcs.	8 x 8 x 16 masonry corner block
Cu. Yds.	Mason's sand
Sacks	Mortar
Sq. Ft.	6 mil. PVC plastic vapor barrier
Cu. Yds.	Pea gravel (crawl space)
Pcs.	8 x 8 x 3/8 steel bearing plates (crawl space)
Pcs.	3-inch steel pipe posts x 7' (crawl space)
Pcs.	4 x 4 x 1/4 steel shims (crawl space)
Pcs.	1/2 x 8 bolts with nuts and washers

Slab Floor

Cu. Yds.	Fill gravel
Cu. Yds	Ready-mix concrete, or

MATERIALS LIST

Cu. Yds.	Sand
Cu. Yds.	Gravel
Sacks	Cement
Sq. Ft.	6 mil. PVC plastic vapor barrier
Sq. Ft.	6 x 6 x #8/8 welded wire reinforcing mesh
Sq. Ft.	Perimeter insulation

Floor Framing and Subfloor

Lin. Ft.	2 x 12 beam (crawl space)
Lin. Ft.	2 x 6 sill
Lin. Ft.	2 x __ joist headers
Pcs.	2 x __ x __ ft. joist
Lin. Ft.	1 x 3 bridging
Sheets	5/8-inch exterior plywood subflooring

Stairs

Pcs.	2 x 10 x 12 ft. stringers
Pcs.	1-1/4 x 10 x 3 ft. 1-1/2 inch treads
Pcs.	1 x 8 x 3 ft. 1-1/2 inch risers
Pc.	1-5/8-inch dia. handrail x 12 ft.

Exterior Walls

Lin. Ft.	2 x 4 plates
Pcs.	2 x 4 x 7 ft. 9 in. studs
Lin. Ft.	2 x 12 headers
Sq. Ft.	1/2-in. plywood filler for headers
Sheets	3/8-in. exterior plywood sheathing, or
Sheets	1/2-in. exterior plywood sheathing

Garage Walls

Lin. Ft.	2 x 4 plates
Pcs.	2 x 4 x 8 ft. 6-5/8 in. studs
Lin. Ft.	2 x 12 headers
Lin. Ft.	2 x 3 beam framing
Sq. Ft.	1/2-in. plywood fillers and beam stiffeners
Sheets	3/8-in. exterior plywood sheathing, or
Sheets	5/8-in. exterior plywood sheathing

Interior Partitions

Lin. Ft.	2 x 4 plates
Lin. Ft.	2 x 6 ceiling backing
Pcs.	2 x 4 x 7 ft. 9 in. studs
Lin. Ft.	2 x 6 headers
Sq. Ft.	1/2-in. plywood filler for headers

Roof Framing and Sheathing

Pcs.	2 x 6 x __ ceiling joists
Pcs.	2 x 6 x __ rafters

202 MATERIALS LIST

Pcs.	2 x 6 x __ rafters
Pcs.	2 x 6 x 3 ft. 9-3/4 in. gable rafters
Lin. Ft.	2 x 8 ridge
Pcs.	2 x 4 x 6 ft. collar beams
Lin. Ft.	2 x 6 fascia backup
Sheets	3/8-in. exterior plywood roof sheathing
Pcs.	4 x 4 x 8 ft. posts (front overhang)

Roofing

Lin. Ft.	Metal drip edging
Sq. Ft.	15-lb. felt
Squares	Asphalt self-sealing shingles, or
Squares	Wood roof shingles or shakes

Exterior Door and Windows

Single Unit	3'4" x 3'2" double hung windows with brick mold
Double Unit	3'4" x 4'6" double hung windows with brick mold
Unit	3' x 6'8" door and frame with weather strip and brick mold
Unit	2'8" x 6'8" door and frame with weather strip and brick mold
Unit	6' x 6'8" gliding door aluminum frame and screen with double glazing and weather stripping
Unit	16' x 7' overhead garage door
Lin. Ft.	2 x 6 garage door frame

Exterior Finish

Lin. Ft.	Brick mold casing
Lin. Ft.	1 x 8 fascia and gable board
Sheets	3/8-in. exterior plywood soffit
Lin. Ft.	Cove molding
Pair	__ in. plastic window shutters
Pair	__ in. plastic window shutters
Pair	80-in. plastic door shutters
Sq. Ft.	Siding

Flashing, Vents, and Gutters

Pcs.	Plastic ridge vents
Pcs.	Under-eave vents
Pcs.	Crawl space vents
Lin. Ft.	Window head flashing
Pcs.	10-ft. box gutter section, aluminum
Pcs.	Slip joint
Pcs.	Endcap
Pcs.	Spike and spacer
Pcs.	Drop outlet
Pcs.	Regular elbow
Pcs.	Rain pipe
Pcs.	Rain pipe strap

MATERIALS LIST

Pcs.	4-in. aluminum vent pipe (exhaust fans and clothes dryer)
Pcs.	4-in. elbow
Pcs.	4-in. eave vent (exhaust fans)
Pcs.	4-in. wall vent (clothes dryer)

Plumbing: Sewer

System	Sewer service to house
Pcs.	4-in. pipe (building drain)
Pcs.	4-in. 45° elbow (building drain)
Pcs.	4-in. Y branch (to secondary soil stack)
Pcs.	4-in. threaded cleanout
Pcs.	4 x 4 x 2 in. Y-branch (to floor drain)
Pcs.	2-in. pipe (to floor drain)
Pcs.	2-in. floor drain and trap (basement)
Pcs.	4 x 3 adapter (to hanging building drain or soil stack)
Pcs.	3-in. pipe (hanging building drain or soil stack)
Pcs.	3-in. 45° elbow (hanging building drain or soil stack)
Pcs.	3-in. Y-branch (hanging building drain)
Pcs.	3-in. threaded cleanout (hanging building drain)
Pcs.	3-in. double sanitary tee with two 1-1/2-in. side outlets (toilets)
Pcs.	3-in. 90° elbow (toilet)
Pcs.	Reducing closet ring flange (toilets)
Pcs.	3 x 1-1/2 sanitary tee reducer (fixture drain)
Pcs.	3-in. roof flashing (soil stack)
Pcs.	3-in. coupling (soil stack)
Pcs.	1-1/2-in. pipe (fixture drains)
Pcs.	1-1/2-in. 90° elbow (fixture drains)
Pcs.	1-1/2-in. P-trap and union (bathtubs)
Pcs.	1-1/2-in. to 1-1/2-in. waste adapter (kitchen sink and bathtubs)
Pcs.	1-1/2-in. to 1-1/4-in. waste adapter (lavatories)
Pints	Cement (plastic pipe)
Spool	Solder (copper pipe)
Jar	Solder flux (copper pipe)
Pcs.	Clamps and gaskets (hubless cast-iron pipe)

Septic System

Cu. Yds.	Gravel
Unit	__ gallon septic tank
Units	Distribution box
Pcs.	4 in. x 10 ft. solid plastic pipe
Pcs.	4 in. x 10 ft. perforated plastic pipe
Pcs.	4-in. coupling
Pcs.	4-in. 1/8 bend
Pcs.	4-in. 1/4 bend
Pcs.	4-in. Y
Pcs.	4-in. cross

MATERIALS LIST

Pcs.	4-in. tee
Pcs.	4 in. x 3 in. plastic adapter

Plumbing: Water

Ft.	3/4 copper tubing—service entrance
Pcs.	3/4 stop and waste valve
Ft.	3/4 rigid copper or CPVC plastic pipe
Pcs.	3/4 tee
Pcs.	3/4 90° elbow
Pcs.	3/4 45° elbow
Pcs.	3/4 coupling
Pcs.	3/4 adapter, copper or plastic pipe to 1/2 threaded pipe
Ft.	1/2 rigid copper or CPVC plastic pipe
Pcs.	1/2 tee
Pcs.	1/2 90° elbow
Pcs.	1/2 45° elbow
Pcs.	1/2 coupling
Pcs.	1/2 union
Pcs.	1/2 pipe cap
Pcs.	1/2 adapter, copper or plastic pipe to 1/2 threaded pipe
Pcs.	1/2 pipe strap
Pcs.	1/2 shutoff valve, outside freeze-proof
Pcs.	1/2 adapter, copper or plastic 3/8 compression (lavatories)
Pint	Cement (plastic pipe)
Spool	Solder (copper pipe)
Jar	Solder paste (copper pipe)
Stick	Pipe dope (threaded pipe)

Well

Unit	Well and pipe
Unit	Pump
Pcs.	Storage tank
Pcs.	Plastic pipe
Pcs.	Plastic elbow
Pcs.	Plastic tee
Pcs.	Plastic connector

Plumbing Fixtures

Unit	Cast-iron bathtub
Unit	Bathtub faucet and shower head
Unit	Bathtub drain
Unit	Toilet
Pcs.	Toilet gasket
Pcs.	3/8-in. toilet supply pipe to wall, with shutoff
Unit	Lavatory, vanity-type
Unit	Lavatory faucet and drain
Pcs.	3/8-in. lavatory supply pipe

Pcs.	Lavatory trap, 1-1/4-in., to wall connection
Unit	Kitchen sink
Unit	Sink faucet with spray and basket strainers
Pcs.	Sink trap, 1-1/2-in., to floor connection
Pcs.	Sink connecting waste, 1-1/2-in.
Unit	Water heater with relief valve
Unit	Double washtub
Unit	Washtub faucet

Plumbing: Gas

Pcs.	1-in. pipe
Pcs.	1-in. union
Pcs.	1-in. tee
Pcs.	1-in. elbow
Pcs.	3/4-in. pipe
Pcs.	3/4-in. union
Pcs.	3/4-in. tee
Pcs.	3/4-in. elbow
Pcs.	3/4-in. cap
Pcs.	3/4-in. gascock
Pcs.	1/2-in. pipe
Pcs.	1/2-in. union
Pcs.	1/2-in. tee
Pcs.	1/2-in. elbow
Pcs.	1/2-in. cap
Pcs.	1/2-in. gascock
Pcs.	3/4-in. brass flexible connector

Heating: Forced Warm Air

Unit	Furnace _____ BTU with plenums
Unit	Furnace humidifier
Unit	Heating–air conditioning thermostat
Ft.	Thermostat wire
Unit	Central air conditioning unit
Pcs.	20 in. x 8 in. x 5 ft. warm air plenum
Pcs.	20 in. x 8 in. x 5 ft. return air plenum
Pcs.	20 in. x 8 in. end caps
Pcs.	Plenum connectors
Pcs.	Plenum hangers
Pcs.	6 in. dia. x 5 ft. round pipe (warm air ducts)
Pcs.	6 in. dia. connector
Pcs.	6 in. top or side take-off (plenum to duct)
Pcs.	6 in. pipe hanger
Pcs.	6 in. volume damper
Pcs.	6 in. 90° elbow
Pcs.	6 in. 90° angle boot (duct to register box)
Pcs.	Floor register box

MATERIALS LIST

Pcs.	Baseboard out-of-wall outlets (warm air)
Sq. Ft.	Gauge sheet metal (return air ducts)
Pcs.	32 in. x 8 in. wall inlet (return air)
Pcs.	Prefabricated chimney with housing
Pcs.	Adapter—6 in. smoke pipe
Pcs.	6 in. dia. x 30 in. smoke pipe
Pcs.	6 in. dia. x 90° elbow
Pcs.	6 in. dia. x 4 in. dia. tee
Pcs.	4 in. dia. x 30 in. smoke pipe (water heater)
Pcs.	4 in. dia. 90° elbow
Boxes	Sheet metal screws
Unit	_____ gallon oil storage tank
Unit	Filter and shutoff valve
Ft.	Copper tubing

Electrical Service Entrance

Pcs.	Entrance head 2-in.
Pcs.	Insulator and bracket 2-in.
Pcs.	Adjustable flashing 2-in.
Pcs.	Conduit support, 2-in.
Pcs.	Conduit 2-in. heavywall x 10 ft.
Pcs.	Eccentric adapter 1-1/4-in. x 2 in.
Pcs.	1-1/4 close nipple
Pcs.	Meter socket
Pcs.	Entrance ell 1-1/4
Pcs.	Conduit 1-1/4-in. heavywall x 10 ft.
Pcs.	Connector 1-1/4-in.
Pcs.	Circuit breaker panel 100 amp. twelve-115V circuits
Pcs.	115V 15-amp. circuit breakers
Pcs.	115V 20-amp. circuit breakers
Pcs.	230V 30-amp. circuit breakers
Pcs.	230V 40-amp. circuit breakers
Pcs.	230V 50-amp. circuit breakers
Ft.	#4 ground wire
Pc.	Ground clamp
Ft.	#3 black insulated wire
Ft.	#3 white insulated wire

Electrical Wiring

Ft.	#12-2 plastic jacket cable (20-amp. circuits)
Ft.	#12-3 plastic jacket cable (20-amp. circuits)
Ft.	#14-2 plastic jacket cable (15-amp. circuits)
Ft.	#__-3 plastic jacket cable (230V circuits)
Lbs.	Staples #10-14 cable
Ft.	#12 black single insulated wire (basement conduit)
Ft.	#12 white single insulated wire (basement conduit)

MATERIALS LIST

Pcs.	Switch or outlet box—new work-type with cable clamps
Pcs.	Octagonal ceiling outlet box with cable clamps
Pcs.	Hanger for ceiling box
Pcs.	Conduit 1/2-in. thin wall x 10 ft. (basement)
Pcs.	Conduit elbow connectors 1/2-in.
Pcs.	Conduit connectors 1/2-in.
Pcs.	Conduit coupling 1/2-in. thin wall
Pcs.	Conduit straps 1/2-in.
Pcs.	Ceiling receptacles with outlet (basement and garage)
Pcs.	1-pole switches quiet type
Pcs.	3-way switches quiet type
Pcs.	Switches, delayed-action (garage)
Pcs.	Switches, dimmer control
Pcs.	Receptacles, heavy-duty grounded (20-amp. circuits)
Pcs.	Receptacles, nongrounded (15-amp. circuits)
Pcs.	Receptacles, outdoor grounded with cover
Pcs.	Receptacles, 230V
Pcs.	Wall plates, single switch
Pcs.	Wall plates, double switch
Pcs.	Wall plates, receptacle
Rolls	Plastic tape
Pcs.	Solder or screw-on solderless connectors

Electrical: Miscellaneous

Unit	Door chime
Pcs.	Push button
Pc.	Doorbell transformer
Ft.	Doorbell wire
Unit	TV antenna system
Unit	Intercom wiring

Electrical Fixtures

Unit	Hall ceiling light fixture
Unit	Kitchen ceiling light fixture
Unit	Kitchen counter light fixture
Unit	Outside light fixture
Unit	Outside pole light
Unit	Bathroom vanity light fixture
Unit	Bathroom ceiling light fixture
Unit	Bathroom electric heater
Unit	Bathroom exhaust fan
Unit	Kitchen exhaust fan
Unit	Attic fan
Unit	Garage door opener
Unit	Intercom

Insulation

Sq. Ft.	6-in. fiberglass ceiling insulation
Sq. Ft.	3-in. fiberglass wall insulation

Drywall

Sq. Ft.	1/2-in. gypsum drywall (sidewalls or ceilings)
Sq. Ft.	3/8-in. gypsum drywall (ceilings)
Lbs.	Drywall joint compound and joint tape
Lin. Ft.	Metal corner bead

Paneling

Sheets	1/4-in. plywood, random-width V-cut prefinished
Lin. Ft.	Base prefinished
Lin. Ft.	Shoe prefinished
Lin. Ft.	Casing prefinished
Lin. Ft.	Cove prefinished
Lin. Ft.	Corner bead prefinished

Ceramic Tile

Sq. Ft.	Unglazed ceramic floor tile
Lin. Ft.	Glazed floor cove and base
Sq. Ft.	Glazed ceramic wall tile
Lin. Ft.	Glazed ceramic wall tile edging
Pcs.	Toilet paper holder
Pcs.	Soap and grab bar
Pcs.	Towel bar
Gallons	Waterproof ceramic tile adhesive (water cleanup)
Lbs.	Latex floor tile grout
Lbs.	White wall tile grout

Interior Doors and Frames

Interior	2'4" x 6'8" x 1-3/8" with frame (baths)
Interior	2'8" x 6'8" x 1-3/8" with frame
Interior	2'8" x 6'8" x 1-3/4" solid core with frame (garage)
Interior	2'0" x 6'8" x 1-3/8" bifold with tracks
Interior	2'6" x 6'8" x 1-3/8" bifold with tracks
Interior	3'0" x 6'8" x 1-3/8" bifold with tracks
Interior	4' x 6'8" x 1-3/8" bifold with tracks
Interior	5' x 6'8" x 1-3/8" bifold with tracks
Interior	6" x 6'8" x 1-3/8" bifold with tracks

Interior Trim

Lin. Ft.	1/2 x 1-3/4 casing
Lin. Ft.	1/2 x 2-1/2 base
Lin. Ft.	1/2 x 3/4 shoe
Lin. Ft.	1/2 x 1-1/4 stop
Lin. Ft.	1 x 12 shelving

Lin. Ft.	1 x 3 hook strip
Lin. Ft.	1 x 2 shelf cleat
Sq. Ft.	1/2-in. interior plywood shelving

Cabinets: Bath

Base	36 x 32 x 21 in. lavatory cabinet
Base	Filler strip
Pc.	__ x 21 lavatory counter and splash with lavatory cutout
Pc.	__ x 21 lavatory counter and splash with lavatory cutout
Pcs.	24 x 36 medicine cabinet
Unit	__ x __ mirror
Pcs.	60 in. x 60 in. aluminum and glass tub enclosure

Cabinets: Kitchen and Laundry

Base	6 x 36 x 24 single cabinet
Base	12 x 36 x 24 single cabinet
Base	18 x 36 x 24 single cabinet
Base	24 x 36 x 24 single cabinet
Base	30 x 36 x 24 double cabinet
Base	36 x 36 x 24 double cabinet
Base	36 x 36 x 24 sink-front cabinet
Base	15 x 36 x 24 drawer cabinet
Base	18 x 36 x 24 drawer cabinet
Wall	30 x 15 x 12 single cabinet
Wall	36 x 15 x 12 single cabinet
Wall	15 x 30 x 12 single cabinet
Wall	18 x 30 x 12 single cabinet
Wall	24 x 30 x 12 single cabinet
Wall	30 x 30 x 12 double cabinet
Wall	36 x 30 x 12 double cabinet
Pc.	__ x 25 counter and splash with sink cutout
Pc.	__ x 25 counter and splash
Pc.	__ x 25 counter and splash

Cabinets: Material

Sheets	1/4 prefinished plywood (doors and exposed ends)
Sheets	1/4 interior plywood or hardboard (backs)
Sheets	3/8-in. interior plywood (door backing)
Sheets	1/2-in. interior plywood (shelves and ends)
Lin. Ft.	1 x 2 pine (shelf supports)
Lin. Ft.	1 x 3 pine (frame)
Lin. Ft.	1 x 4 pine (frame)
Lin. Ft.	1 x 12 (base cabinet shelf)
Sets	Drawer guide hardware
Pairs	3/8-in. inset cabinet hinges
Pcs.	Magnet door catches
Pcs.	Door pulls

210 MATERIALS LIST

Pcs.	Drawer pulls
Pcs.	Counter and splash (see above), or
Sheets	3/4-in. interior plywood (tops)
Sheets	1/16-in. formica laminate counter top

Appliances

Unit	36-in. refrigerator
Unit	24-in. dishwasher
Unit	Garbage disposal
Unit	30-in. clothes washer
Unit	30-in. clothes dryer
Unit	30-in. range and oven
Unit	30-in. range hood

Finish Hardware

Front	Exterior door lock set
Rear	Exterior door lock set
Interior	Privacy lock set (bathrooms, master bedroom)
Interior	Door lock set
Pairs	3-1/2 x 3-1/2 butt hinges (interior doors)
Pairs	4 x 4 butt hinges (exterior doors) Sets Bifold door tracks and hardware
Pcs.	Door stops
Pcs.	1-in. electrical conduit (for closet poles)
Pairs	Closet pole brackets
Sets	Window locks and pulls
Pcs.	30-in. exterior door thresholds
Pc.	Mailbox
Sets	Stair handrail brackets

Nails

Lbs.	30d common nails (framing)
Lbs.	20d common nails (framing)
Lbs.	16d common nails (framing)
Lbs.	10d common nails (framing)
Lbs.	Roofing or shingle
Lbs.	Drywall nails
Lbs.	8d finish nails (interior trim)
Lbs.	6d finish nails (interior trim)
Lbs.	4d finish nails (interior trim)
Lbs.	Siding nails (coated)
Lbs.	Paneling nails (coated)
Lbs.	8d masonry nails
Lbs.	8-in. joist hangers (framing)
Lbs.	10-in. joist hangers (framing)

Finish Floors

Sq. Ft.	3/8-in. interior plywood underlayment (basement and c.s.)
Sq. Ft.	Flooring felt
Sq. Yd.	Kitchen–family room flooring
Sq. Yd.	Carpeting and pad

Paint and Wallpaper

Sheets	Sandpaper
Qt.	Nail hole patching paste
Tubes	Exterior caulking
Qts.	Interior wood primer
Gallons	Exterior wood primer
Qts.	Metal primer
Gallons	Exterior siding paint
Gallons	Exterior siding stain
Gallons	Exterior trim paint
Qts.	Interior trim paint, semigloss
Qts.	Interior trim stain
Qts.	Interior trim varnish
Gallons	Interior wall paint, semigloss (baths and kitchen)
Gallons	Interior wall paint, flat
Gallons	Porch and deck enamel
Qts.	Metal enamel
Rolls	Wallpaper
Rolls	Border
Pkgs.	Wallpaper paste
Sticks	Putty stick for stained wood

Concrete Garage, Drive, Walks

Cu. Yds.	Ready-mix concrete, or
Cu. Yds.	Sand
Cu. Yds.	Gravel
Sacks	Cement
Sq. Ft.	6 x 6 x #8/8 welded wire reinforcing mesh
Cu. Yds.	Fill gravel
Ft.	Concrete reinforcing rods

Brick

Thousand	Common or face brick
Cu. Yds	Mason's sand
Sacks	Mortar
Pcs.	3-1/2 x 3-1/2 x 3/8 angle iron x __
Pcs.	3-1/2 x 3-1/2 x 3/8 angle iron x __
Pcs.	3-1/2 x 3-1/2 x 3/8 angle iron x __
Pcs.	Vents (crawl space)
Pcs.	22 gauge galvanized brick anchors
Pcs.	__ stone window sill

MATERIALS LIST

Pcs.	__ stone window sill
Pcs.	__ stone door sill
Pcs.	__ stone door sill
Gallons	Muriatic acid (for cleaning brick)

Fireplace
Unit	Prefabricated fireplace
Unit	Prefabricated fireplace chimney

Index

Acoustical ceiling tiles, 151
Air conditioning, 126–27, 167
    drains for, 102
    specifications for, 198
Ampere, 128
Appliances, 166–67, 210
    wiring for 230-volt, 147
Attorneys, 4, 193

Basement, 6
    building drains for, 93, 95, 98
    excavation for, 23–24
    floor, 168–70, 200
    floor framing, 48
    foundation for, 29–36, 199–200
    furnace in, 121
    waterproofing, 38–39
    water supply for, 107–10
Bathrooms, 6
    elevations of, 158
Bathtubs, 113
    drains for, 97
    water supply for, 108–9
Batter boards, 19–20
Bedrooms, 5–6
Board and batten siding, 89–90
Brick flooring, 163
Bricks
    laying of, 171–73
    materials for, 211
Brick veneer
    exterior sheathing for, 55–56
    insulation for, 149
Building codes, 3–4, 14, 94, 193
Building drains, 92–96, 98–99
Building layout, 20
Building materials, purchasing, 17–18.
    See also Materials list
BX. *See* Cables, flexible armored

Cabinets, 153–58, 209
Cables
    flexible armored (BX), 130, 137, 143
    nonmetallic-sheathed (Romex), 130, 137, 143
Carpentry and millwork, specifications for, 194–95
Carpeting, 161
Carport, 7
Ceiling finish, 150–52
Cement, for footings, 28
Center beam, 50
Chimney, prefabricated, 120
Circuit, 128
Closets, 6
    doors for, 83
Clothes washer, 166–67
    drains for, 102
Concrete and masonry, specifications for, 194
Concrete block foundation, 29–35
Concrete flatwork, 168–70, 211
Conduit, 130, 137–38, 143
Copper pipe, 94, 107
Corner construction, 57
Crawl spaces
    building drain for, 93, 95, 98
    excavation for, 23–24
    floor framing, 48–49
    furnace in, 121
    materials list for, 200
    waterproofing, 38
Currents, electrical, 129

Dining room, 6
Dishwasher, 166
    drains for, 102
Disposal field for septic system, 104–6
Door framing, interior, 60

INDEX

Doors, 83–85
  materials for, 202, 208
  and trim, 160
Downspouts, 88
Drainage plumbing, 92–102
Driveways, 168
DWV (drainage-waste-vent) plastic pipe, 94

Electrical installation, 134–47
Electrical service, 128–33, 206
Electrical switches and receptacles, installation of, 139–40, 144–46, 147
Electrical symbols used on floor plans, 131
Electrical system
  materials for, 206–7
  specifications for, 197–98
Electric boxes, installation of, 136–37
Electric heating, 117, 118
Electric meter socket, wiring of, 142
Elevation drawings, 11
Entry hall, 6
Excavation, 23–25
Excavation layout, 21

Family room, 6
Financing, obtaining, 12–13
Finish floors, 161–63, 210
Fireplaces, 174–75, 212
Flashings, 86
  materials for, 202
Floor finish, 161–63
  specifications for, 195–96
Floor framing, 48–49
  materials for, 201
  plan, 52
Floor opening, 51
Floor plan, 11
Footings, 23–25, 26–28
  materials for, 199
Forced warm air heat, 117, 121–24
Foundation, 29–37
  concrete block, 29–35
  crawl space, 37
  for basements, 29, 36
  for slab floors, 40–43
  materials for, 199–200
  pier, 44–47
  plans, 11, 36–37, 42, 46
  poured concrete, 30–31
  waterproofing, 38–39
Furnace
  for forced warm air heating, 121–24
  gas, 115

Garage, 7
  beam, 63
  excavation for, 25
  floor, 168–70
  insulation of, 148
  interior walls of, 152
  materials for, 200, 201
  wall framing for, 56–57, 62
Gas
  bottled, 116, 118
  for heating, 118
  plumbing for, 115–16, 205
Gas meter, 115
Grading land, 176–77
Grounded receptacles, 144, 147
Gutters, 87–88
  materials for, 202
Gypsum drywall, 150–51

Hardware, finish, 210
Hardwood floors, 161, 162–63
Heating
  fuel for, 117–18
  materials for, 205–6
  specifications for, 198
  types of, 117–18
Home Builders Plans, 8
Home Planners, Inc., 8
Hot water baseboard heat, 117, 118, 119–20
House plans, 8–11
Hubbed cast-iron pipe, 95
Hubless cast-iron pipe, 94–95, 99

Incinerators, 167
Insulation, 148–49
  materials for, 207–8
Insurance, 16, 193
Interior partitions, 56
  framing details, 64
Interior trim, 159–60

Kilowatt-hour, 129
Kitchen, 6
  elevations for, 158
Kitchen sink, 114, 166
  drains for, 97, 102
  water supply for, 109–10

Land, purchasing, 3–4
Lavatory, 113–14
  drains for, 97
  water supply for, 108–9
Lawn, 176–77
Linoleum, 161–62
Lot, inspection of, 3

INDEX 215

Lumber sizes, 178–79

Materials, building. *See* Building materials
Materials list, 199–212
Molding, shoe, 159
Motor, 29–30, 172–73

Nailing surface for ceiling material, 72
Nails and screws, 180, 210
National Electrical Code, 129
National Plumbing Code, 94, 107

Outlets, electrical, 131–33

Painting, 211
   exterior, 164
   interior, 164
Paneling, 151, 208
Pier foundation, 44–47
   building drain for, 93–94, 95, 98
   water supply for, 110
Pipes, types of, 94–95, 107–8
Planning a house, 5–7
Plans, house. *See* House plans
Plastic pipe, 94, 107
Plot planning, 5, 14–16
Plot plans, 14–16
Plumbing
   drainage, 92–102
   materials for, 203–5
   specifications for, 196–97
   water supply, 107–110
Plumbing fixtures, 113–14
Plumbing for gas, 115–16

Rafters, roof, 66–75
   details, 68–69
Raised-ranch house, 7
Romex. *See* Cables, nonmetallic-sheathed
Roof
   framing, 66–75
   gable end detail, 74
   materials for, 201–2
   sheathing, 73, 75
   shingles, 76–78
Roofing, 76–78
   materials for, 202
Rooms, number and use of, 5–7

Screws and nails, 180
Seeding, 176–77
Septic system, 103–5
   materials for, 203–4
Service entrance, electrical, 134–36, 141
Service entrance panel, 135, 142

Sheathing, exterior, 55–56
   details, 65
   roof, 75
Sheet metal, specifications for, 198
Shingles
   for exterior siding, 91
   for roofing, 76–78
Shoe molding, 159
Siding, exterior, 89–91
Sinks
   drains for, 97, 102
   water supply for, 108–110
Site, preparation of, 19–22
Slab floors, 40–43
   building drain for, 93, 95, 98
   excavation for, 23–24
   heating systems for, 121, 125
   materials for, 200–1
   water supply for, 110
Slate flooring, 162
Sodding, 176
Soil stack, 92, 93, 95, 96
Specifications, 193–98
Split-level houses, 7
Stair openings, 53
Stairs, materials for, 201
Steel pipes, 107–8
Storage, outdoor, 7
Storm doors, 85
Storm windows, 79
Subflooring, 49–53
   materials for, 201
Sump pump, 93
Switch wiring, 139–40, 144–46, 147

Termite shield, 48
Tiles, ceramic, 151–52, 162, 208
Toilets, 113–14
   basement, 93
   drains for, 96–97, 100
   dry, 106
   water supply for, 108–9
Tools, 181–82
Trees, planting, 176
Trench forms, 26–28
Trim
   details, 179
   interior, 159–60, 208
Two-story houses, 7

Underlayment for flooring, 161
Utilities, 4
Utility room, 6

Vanity tops, 153
Vapor barrier, 149

216 INDEX

Vents, 86–88, 99
    materials for, 202
Vinyl floor tiles, 162
Volt, 128

Wall finish, specifications for, 196
Wall framing
    details, 61
    exterior, 54–55
    garage, 56, 62, 63
    materials for, 201
Wall intersection, 58
Wallpapering, 164–65, 211
Walls, interior finish, 150–52
    materials for, 201
Water heater, 108, 113, 115
Water meter, 107, 108

Waterproofing foundations, 38–39
Water softener, drains for, 102
Water supply, plumbing for, 107–10
Watt, 129
Wells, 111–12
    materials for, 204
Window framing, 59
Windows, 79–82
    materials for, 202
Window wells, 31
Wire sizes, electrical, 138–39
Wires, joining electrical, 144
Wiring circuits, typical house, 132
Wiring systems, 130, 137–40, 143, 147
    materials for, 206–7
Wood forms, 28
Wood posts, 44

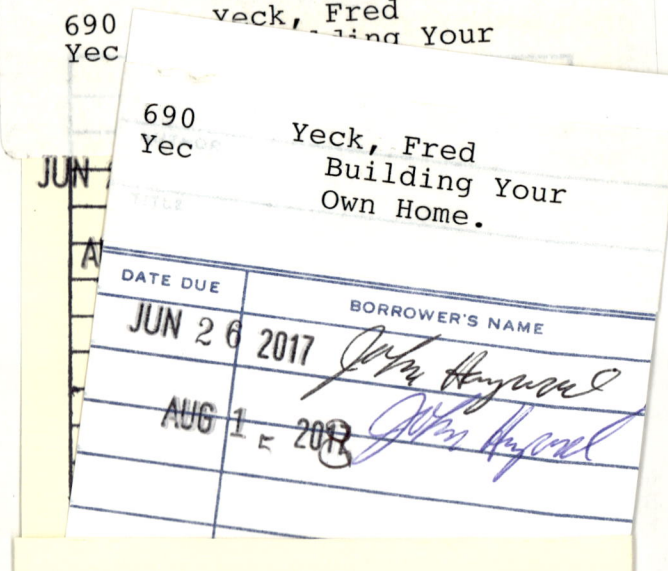